Contents

International Food Data Bases and Information Exchange

World Review of Nutrition and Dietetics

Vol. 68

Basel · München · Paris · London · New York · New Delhi · Bangkok · Singapore · Tokyo · Sydney

International Food Data Bases and Information Exchange

Concepts, Principles and Designs

Volume Editors

Artemis P. Simopoulos

The Center for Genetics, Nutrition and Health, American Association for World Health, Washington, D.C.

Ritva R. Butrum

Rockville, Md.

22 figures and 21 tables, 1992

Basel · München · Paris · London · New York · New Delhi · Bangkok · Singapore · Tokyo · Sydney

World Review of Nutrition and Dietetics

Library of Congress Cataloging-in-Publication Data
International food data bases and information exchange: concepts, principles, and designs / volume editors, Artemis P. Simopoulos, Ritva R. Butrum.
(World review of nutrition and dietetics; vol. 68)
Includes bibliographical references and index.
1. Food – Composition – Data bases. I. Simopoulos, Artemis P., 1933– . II. Butrum, Ritva Rauanheimo, 1927– . III. Series.
[DNLM: 1. Food Analysis. 2. Information Storage and Retrieval. 3. Nutritive Value.]
ISBN 3-8055-5480-X (alk. paper)

Bibliographic Indices
This publication is listed in bibliographic services, including Current Contents® and Index Medicus.

Drug Dosage
The authors and the publisher have exerted every effort to ensure that drug selection and dosage set forth in this text are in accord with current recommendations and practice at the time of publication. However, in view of ongoing research, changes in government regulations, and the constant flow of information relating to drug therapy and drug reactions, the reader is urged to check the package insert for each drug for any change in indications and dosage and for added warnings and precautions. This is particularly important when the recommended agent is a new and/or infrequently employed drug.

Printed in Switzerland on acid-free paper by Thür AG Offsetdruck, Pratteln
ISBN 3-8055-5480-X

Preface

Over the past 15 years the scientific basis of nutrition has expanded extensively, particularly on the role of diet in the development of chronic diseases such as cardiovascular disease, hypertension, autoimmune diseases and disorders, and some forms of cancer. Our understanding of the role of nutrients in gene expression makes it mandatory that we have precise information of the composition of the food supply.

The Food and Agriculture Organization (FAO) of the United Nations published food composition tables in 1949 and 1954 for international use. The tables were based on analytical data from different parts of the world and were intended to be used in countries that had no food composition data for the development of Food Balance Sheets, based on the comparison of the nutrient content of food supplies in the different countries.

Information about storage and analytical data about animal nutrition has preceded information on human nutrition. Thus the extensive work and progress achieved by the International Network of Feed Information (INFIC) illustrates that the interest for an international interchange of data on feedstuffs preceded that for foodstuffs, the International Network of Food Data Systems (INFOODS). INFIC had already proposed an international coding system for feeds in the 1960s whereas INFOODS came into being in 1984–1985 following the meeting at Bellagio, Italy, in 1983. Today more than 150 food composition tables are listed in the INFOODS directory of food composition tables. The important role played by a number of organizations who met at Bellagio included representatives from FAO of the United Nations, the International Union of Nutritional Sciences (IUNS), the International Union of Food Sciences and Technology (IUFST), the US Department of Agriculture (USDA), the Nutrition Coordinating Committee (NCC) of the National Institutes of Health (NIH), and the National Cancer Institute (NCI) of NIH. The last two organizations were represented by the editors of this volume. A.P. Simopoulos represented the Nutrition Coordinating Committee of the NIH and R. Butrum the National Cancer Institute.

This volume begins with an 'Overview of International Food Composition Data Bases' by Dr. Butrum. Dr. Butrum discusses the needs of users of food composition data bases. The users include international researchers interested in identifying specific dietary problems worldwide, international agencies that monitor the dietary needs of a specific population for the provision of food supplies to those in need. Often it is necessary to meet specific nutrient needs. It is therefore essential to have information on the composition of the foods available in the country and of the foods to be distributed. Food industries competing in an international market need to know the components of the local food supply and their own foods' nutrient composition. Another section in this paper is on the development of food description systems and the author predicts that as the food supply grows and the role of diet in normal growth and development and in health and disease is further elucidated, the number of food data bases as well as their significant role in research and public policy will increase.

The second paper 'International Food Database: Conceptual Design' by S. Srivastava and R. Butrum describes the efforts of the NCI, the issues involved in the development of International Food Data Bases such as copyright infringement, technical problems, description of foods, etc. The paper describes extensively a food description language and data base design; the software and the need to choose a data base management system; and presents a conceptual model of the data base and emphasizes that this model is intended for a wide spectrum of users, researchers performing consumption studies, epidemiologists, clinical investigators, etc. and describes international cooperation among various scientists.

The next paper by D.A.T. Southgate and H. Greenfield discusses the 'Principles for the Preparation of Nutritional Data Bases and Food Composition Tables'. Detailed information on nutrients and other constituents to meet most data base user needs, and the specifics of the data base and criteria for scrutiny of data is presented in table format. The authors state that food data bases are in essence 'tools' for a wide range of nutritional tasks from the management of metabolic disease states to research, particularly in lipid research, and teaching; in the assessment of the adequacy of the food supply, dietary intake of populations, and in the development of food and nutrition policies by governments; and in developed countries nutritional or food data bases form the basis for the nutritional labeling of foods and food products.

The fourth paper 'Validated Data Banks on Food Composition: Concepts for Modeling Information' by M. Feinberg et al. discusses food composition data bases and presents ways of computing methodology and its successful application to structuring complex chemical information. The authors state that to this day none of the existing food nutrient banks is accessible to online users, and describe the REGAL system, which makes possible the storage of as many synonyms as necessary and the publication of multilingual tables. The name of food does not reflect its chemical composition, for example using a dictionary, 'bread' translates into 'pain' (French) or 'brot' (German), but this does not mean that French or German bread is made the same way as English or American bread. To provide a scientific description for food, a Committee on Data for Science and Technology (CODATA) Task Group on a Systematic Nomenclature for Foods in Numeric Data Banks has been formed to study the descriptive codification of foods. The CODATA is an interdisciplinary Scientific Committee of the International Council of Scientific Unions (ICSU) that seeks to improve the quality, reliability, management, and accessibility of data of importance to all fields of science and technology. The authors comment on the EUROCODE and LanguaL systems. Modifications and improvements are being made by the CODATA. LanguaL's coding system permits modification even in existing data banks. Globally, LanguaL remains a good starting point for development of a truly international and flexible faceted thesaurus.

The paper by Thomas C. Hendricks provides a detailed description of LanguaL under the title 'LanguaL. An Automated Method for Describing, Capturing and Retrieving Data about Food'.

The next paper by A. Møller is a report on the 'NORFOODS Computer Group'. NORFOODS is the Nordic group of projects concerning food composition tables and data banks. The countries involved are: Denmark, Finland, Iceland, Norway and Sweden. NORFOODS describes the first systematic interchange of data of food composition tables in the world, and it suggests a minimum standard run on data format by electronic data transfer of food data among the Nordic countries. It may be applicable for data transfer among other countries that have developed food composition data banks.

In 'Past and Present Activities in Food Composition Tables in Latin America and the Caribbean Islands' R. Bressani and M. Flores discuss the development of national, Latin American and Caribbean Islands (LATINFOODS) food composition tables, their status, and future prospects.

The last paper by J.H. Dobrzycki and M. Los-Kuczera on 'Food Composition Tables and Food Composition Analysis in East Europe' describes the published East European Food Composition Tables (EEFC) currently in use and considers the possibility of using these tables in an international computerized information system. The countries represented are Czechoslovakia, Hungary, Poland, Bulgaria, the (previous) German Democratic Republic (GDR) and Yugoslavia. Because these tables are written in the country's language, except for Czechoslovakia that has an English introduction and a Czech-English, English-Czech index of the products, the inclusion of Latin names for plants, fish, etc., should make their use possible. However, differences in the classification of food groups hinders the comparison of any tables. These and other problems are discussed. The authors however are optimistic that the improvements and dramatic increase in the use of computers 'along with their readiness to implement East European Food Composition Tables into the regional food data banks, will be the impetus for building a unified international food data bank'.

In any quantitative study of human nutrition, information on the nutrient composition of foods is essential in order to calculate dietary intake, to assess the nutritional adequacy of diets, formulate dietary modifications, to exchange nutrient composition data nationally and internationally, to evaluate the role of dietary factors in diseases and conditions under investigation, for food labeling, and in the overall development of nutrition policy. The names and descriptive terms for the foods must be easily understood and unambiguous. The lack of a food description language and differences in the format of reporting units of measurement and methods of analyses have seriously limited the use of food tables for comparing nutrient contents of particular foods. These problems, however, are being tended to by a number of investigators around the world as this volume attests to. At present the International Food Database is in the developmental phase.

Information about international food data bases is essential for everyone working in human nutrition such as clinical investigators, epidemiologists, nutritionists, dietitians, food technologists, computer scientists, obstetricians, pediatricians, internists, general practitioners, and in general, scientists in government, industry and academia, as well as policymakers, lawyers, informed consumers, and those involved in food marketing and food regulation.

Artemis P. Simopoulos, MD

Simopoulos AP, Butrum RR (eds): International Food Data Bases and Information Exchange. World Rev Nutr Diet. Basel, Karger, 1992, vol 68, pp 1–14

Overview of International Food Composition Data Bases

Ritva R. Butrum

Rockville, Md., USA

Contents

Introduction

Food composition data bases are widely used to evaluate the nutrient content of the food supply and specific diets. For this reason, they constitute an integral part of the knowledge required to understand the roles played by the nutritional environment in human health and disease prevention. Recent emphasis on studies assessing the relationships between nutrition and chronic diseases such as cancer, coronary heart disease, diabetes, and hypertension for various population groups has stimulated interest in standardized detailed food composition data. For example, epidemiological studies are largely dependent on food composition data bases to evaluate relationships between food consumption and risk assessment.

These data bases are used because of the cost and impracticality of obtaining and analyzing foods from a large number of free-living subjects required for such studies [1]. However, these studies are limited in assessing dietary risk and clarifying etiologic relationships because detailed information on both the nutrient and nonnutrient content of many foods is lacking.

Developing accurate and reliable food composition data bases is essential because both research and public policy in the area of nutrition and health requires standardized accountable information. This chapter outlines the evolution of food composition data bases, including a description of the numerous users and needs for data bases, historical perspectives, development of various vocabulary systems, and current National Cancer Institute (NCI) initiatives in this area.

Food Composition Data Bases: Users and Their Needs

The quantity and complexity of available foods on the market is increasing. While the average US supermarket carried 867 items in 1928, there were 24,000 items available in 1988 [2]. In addition, the number of constituents thought to be biologically important for disease prevention are increasing. As the roles of nutritive and nonnutritive components in foods are clarified, the uses for food composition data bases will expand. Current users for data bases, ranging in scope from multinational and international organizations to the individual consumer, are described below.

Generally, food composition data bases are used in international research projects to identify specific dietary problems worldwide by analyzing regional food consumption patterns. International agencies monitoring worldwide populations provide large quantities of food to nutrition-deficient countries around the world, often to meet specific nutrient needs. To accomplish this, agencies need to know the constituents of food stuffs that are available around the world and at the same time the relevant nutrient situation within each country so that they can match the food supply and availability to human physiologic needs [3]. In addition, food industries competing on the international market need to know the components of the local food supplies with which they are competing and the nutrient contents of their own products.

Nationally, governments fund studies of food composition and health issues, including the monitoring of food imports and exports, for both nutritional content and compliance with health and safety regulations. They assess the nutritional status of their people by estimating nutrient and nonnutrient consumption from surveys of food disappearance data, household food purchases, and records of dietary intake, and by comparing actual consumption with established requirements. Food composition data bases are used to establish food identities, to control substitution of ingredients, to check the validity of advertising, and to formulate labeling laws. National feeding programs, such as those found in schools or the military, require updated information on the nutrient content of foods so that nutritionally adequate diets can be formulated [3].

Public programs and private organizations on a local level, responsible for feeding thousands daily, must comply with nutrition guidelines. Local organizations, such as hospitals interested in nutrition and health, rely on food composition data for patient and public education. Universities with teaching and research programs in nutrition also require reliable and current food composition data bases. In response to public demand for improved nutritional quality of food products, the food industry uses food composition data as a basis to change and regulate product content. Food data bases are used to ensure that new product formulations adhere to nutritional and safety standards.

On an individual basis, food data are used by physicians and dietitians to better understand the relationship between diet and disease, to evaluate patient nutritional status, and to devise eating programs to foster health promotion. Food composition data bases are used to assess dietary intake of subjects participating in clinical trials related to chronic diseases such as cancer and heart disease. In recent years, the public has developed an interest in nutrition information that has led to the development of food guides and other materials for determining intake of calories and other nutrients for facilitating food choices. Moreover, one of the most extensive uses of food composition data by the individual has become the shopper scanning the ingredient list, nutrient contents, and percentages of requirements fulfilled on the labels of packaged foods [3].

Historical Perspective

Over the years, the complexity of users' needs has resulted in the development of numerous data bases containing data of varying reliability. These tables contain average values that do not reflect variability among samples of the same food, nor do they reflect differences in geographic location over time. Limited data exist for specific classes of foods such as commercially prepared convenience foods and fast foods. In addition, biases in food composition data may result when imprecise or inappropriate analytical methods are used or when a food is incorrectly identified [4]. Further, many of these data bases are incompatible with one another and are often difficult to assess, leading to further diversity of food composition data resources. Moreover, these problems become more acute as new foods and new methods of production and storage are continually developed and as international markets and trade in foods expand.

To address these concerns, the author and Drs. Sorenson and Seltzer drafted a position paper in 1981 to assess the feasibility of establishing an international food data base system. Based on this paper, representatives of the food industry, international organizations, universities, and government agencies attended a meeting in June 1982 in Washington, D.C., to evaluate the issues related to quality nutrient composition data and seek support from the Rockefeller Foundation. Discussions at this Washington, D.C. meeting centered around the need for and current limitations of food composition data bases, particularly in an international context. It was proposed that an organization be established to promote international participation and cooperation in the acquisition and interchange of quality data on the nutrient composition of foods, beverages, and their ingredients in forms appropriate to meet the needs of government agencies; nutrition scientists; health and agricultural professionals; policymakers and policy planners; food producers, processors, and retailers; and consumers. The name of the organization, INFOODS (International Network of Food Data Systems), was proposed by Dr. Joseph Street of Utah State University.

In January 1983, the Food, Nutrition, and Poverty Subprogramme of the United Nations University, with the support of various US government agencies, private foundations, and the food industry, sponsored the Bellagio Conference. Held at the Rockefeller Conference and Study Center in Bellagio, Italy, this conference included representatives from the Food and Agriculture Organization (FAO), the World Health Organization (WHO), the International Union of Nutritional Sciences, the International Union of Food Science and Technology, the US Department of Agriculture, Nutrition Coordinating Committee (NIH), and the National Cancer Institute of NIH. The goal of the conference was to explore and develop approaches with a view to defining an overall strategy and course of action that would promote establishment of a standardized, high quality, readily accessible international food data system [5]. The conference participants agreed that to successfully fulfill this mission, INFOODS would need to develop four task areas: (1) a network, not only of data bases, but also of people and organizations; (2) standards and guidelines relative to data gathering, data storage and interchange, and data usage; (3) a mechanism to handle information collection and dissemination, coordination, identification of needs, and resource allocation, and (4) an international journal devoted to food composition studies [3].

Following the Bellagio Conference, the Diet and Cancer Branch (DCB), NCI, sponsored a project to accomplish the following areas of work:

- An international survey of users and users' requirements within government, industry, and universities
- Establishment of guidelines for acceptable data in terms of sampling for food analysis, preparation of samples, and analytical techniques
- A survey and assessment of existing food data systems
- A documentation system for foods and food components
- Development of a data base management system responsive to users' needs
- Development of criteria for determining values from existing data bases that meet standards for inclusion in the new data system
- Design of input forms to be used in soliciting data from different sources
- Establishment of liaison committees with international organizations and representatives from governments, academia, and the food industry

To accomplish these objectives, INFOODS proposed a separate working and steering committee for four areas. These committees included the Users and Needs committee, Data Quality committee, Terminology and Nomenclature committee, and Information Systems committee.

The major activity of the User and Needs committee was the organization of a conference in March 1985 to present and discuss the general topics of the uses and needs of food composition data. A summary of this conference and the principal papers presented were edited by INFOODS and published in November 1987 as *Food Composition Data: A Users' Perspective* (Rand, Windham, Wyse, Young, eds). In addition, the International Directory of Food Composition Data was prepared by Dr. William Rand and staff at MIT. This effort involved contacting many countries of the world and collecting a nonannotated directory of food composition tables currently in use.

Although a regional food data organization in Europe, EUROFOODS, had already been formed to examine the state of data in the region, other countries and regions were in need of similar organizations. In collaboration with INFOODS, several groups were formed, including NORFOODS, ASIAFOODS, LATINFOODS, OCEANIAFOODS, and AFROFOODS. The fundamental objective of these groups is the sharing of scientific knowledge and international collaboration relating to food composition data. National committees and effective channels of communication between countries in these groups will determine specific needs, mobilize available resources, and establish an environment favorable to the development of quality food composition data.

The usefulness of any data base depends on the quality of the data used to construct it. To foster the advancement of quality data bases, Dr. D.A.T. Southgate, Chairman of the Data Quality Committee, in collaboration with Dr. Heather Greenfield, is revising and expanding a document that covers in depth the procedures for collecting and analyzing food samples within the full context of developing food composition data bases. This book should prove invaluable to those developing food data bases everywhere. It may also improve the likelihood that data coming from new analyses will be usable for comparative studies. More details concerning this work can be found in the chapter by Dr. Southgate [pp. 27–48].

The documentation of data involving the identification of foods, nutrients, and data for intelligible interchange between systems is essential for food composition data base development. Therefore, the Terminology and Nomenclature committee determined that the critical food nomenclature issue was the establishment of guidelines to describe the nature and character of the food. The committee also recommended the development of an international coding system that focuses on the utilization and supplementation of existing systems rather than replacing them. Abbreviated 'tagnames' have been associated with each of the nutrients and nutrient determination methods so they can be easily recorded and used in interchange. A series of reports was developed, including some generic words for food identification, depicting food-structured descriptions rather than assigning specific names.

Of particular concern in the development of food composition data bases is the issue of missing data. Although it is common that there are not good reliable methods for assaying particular components or identifying appropriate data sources, many of the data do not exist simply because a particular component was not of interest at the time. Users with specific needs have two options: they can generate the data themselves, or they can estimate (impute) the missing values from known data on similar foods and components. The first option requires resources that users rarely have available, while the second requires clear and well-defined rules for estimation, which do not currently exist. In October 1986, INFOODS sponsored a conference in Washington, D.C., to discuss issues related to handling missing values in food composition data bases.

Very early in the project, it was tentatively concluded that a single data base incorporating all of the food composition data in the world was impractical and inappropriate. The relatively high frequency with which values are changed and new international tables produced, in comparison with the likely frequency of need for new copies of foreign data, makes it difficult and expensive to maintain a single, up-to-date worldwide data base and to verify that it was up to date. Consequently, INFOODS developed an interchange system to support international data in a 'distributed data base management' environment rather than a single data base. An advantage to this system is that while it inherently encourages data base compilers to maintain more extensive and higher quality documentation about their data than has been done in the past, it does not require data base maintainers to change the procedures or organizations of the data bases. Using the international standard markup language, prototype codes were developed that allow interaction between the formats of several conventional data bases. The interchange format and codes have been tested with a few national data bases [6].

Within this interchange system, it may be possible to transfer data between different data bases. However, the system lacks specificity in regard to important issues such as food identification, origin, and quality of data. Standard procedures are necessary for representing and processing data in various systems so that ambiguities do not exist in either the identification of the foods or interpretation of the data. Current initiatives in this area are discussed later in this chapter.

In general, INFOODS has been successful in raising important issues relevant to the area of food composition such as identifying omissions, maintaining and updating data bases, and regarding food data in more scientific terms. Other considerations include the importance of statistically based sampling strategies, analytical methods suitable for the component in question, and available resources. Efforts have included making some important information and practices related to these areas more generally known and available because one of the difficulties of the field has been that procedures developed in one country or site may be unknown to others. Through a quarterly newsletter and an international journal, *Journal of Food Composition and Analysis,* INFOODS has fostered communication between the various scientific bodies involved in this area.

Development of Food Description Systems

Over the years it has become increasingly important for investigators to exchange food composition data on both a national and international level. In the past, items of human foods were classified and named, and their characteristics and origin were described independently by individual data base managers. For identifying foods, scientists need a single systematic convention that is sufficiently flexible to accommodate the variety of potential users for food data and that will forward the flow of information [7].

Since the early 1970s, DCB, in conjunction with the Food and Drug Administration (FDA) and other organizations, has established systems where names and descriptive terms for the foods, nutrients, and other constituents found in these data are precise, easily understood, and unambiguous. Although in theory it may appear simple to establish or interpret food names and descriptors, in practice, ambiguity and lack of specificity may be problems. Often regional names for a food are used, and that name can refer to different foods in different areas. For example, battikha in Arabic means watermelon in the Middle East and melon in North Africa [8]. This can affect dietary intake assessments of the population as different species of foods contain different constituents.

International Network of Feed Information Centers

Efficient storage, retrieval, and processing of data in computerized food-related data bases require a controlled vocabulary with a thesaurus that clearly defines the vocabulary terms [9]. One such vocabulary system, the International Feed Databank System, has been developed by the International Network of Feed Information Centers (INFIC). INFIC adopted a computerized system for describing and recording information about feeds, enabling the descriptive information and numerical data associated with various feeds to be stored, summarized, retrieved, and printed in various formats. Online information is also available for developing animal diets that will yield maximum profit [9].

Factored Food Vocabulary (FFV)

Although the INFIC system is international in scope, it describes feeds rather than foods. The Center for Food Safety and Applied Nutrition of the FDA has developed a standardized language for describing foods, specifically for classifying food products for information retrieval. The structure of the FFV is based on two main ideas: (1) a food product can be described by a combination of several characteristics, each of which may serve as a retrieval term or descriptor for the food product, and (2) the characteristics can be brought together in a meaningful classification relating them to one another [10]. The FFV was developed to facilitate retrieval and aggregation of information about foods from various data files. The 13 factors used to describe foods are listed in table 1. The terms for each factor may be retrieved and/or aggregated in various combinations, and the sequence of the factors was chosen to facilitate the writing and reading of food product descriptions and to aid in the comprehension of the vocabulary [10].

For each factor there is a hierarchy of terms from broader to narrower. The indexer selects from each factor the specific term that most accurately describes the food. For example, one could search the broad term 'nutrient or dietary supplement added', or more narrowly 'vitamin added', and still more specifically, 'vitamin A added'. This hierarchical arrangement also displays the vocabulary in a logical way to facilitate indexing and retrieval. It provides definitions and cross-indexing and gives examples of how the terms should be used in indexing.

A data file coded with the FFV can be sorted and displayed by any of the factor terms or by various combinations of factor terms. Foods in various coded data files can be matched to compare nutrient data or to aggregate various kinds of food information. For example, information from a food consumption data file can be combined with data from a nutrient composition file by linking the code strings for each food [9]. Efficient and accurate aggregation of information concerning foods can be obtained using this system. With this system, factor terms can be changed, added, or deleted as needed, thus offering flexibility with specificity.

In a collaborative effort to validate the system, the FDA and NCI have used the FFV to factor the foods in the updated version of the United States Department of Agriculture Handbook No. 8. The Handbook No. 8 described by the FFV provides much more detailed information about the foods and allows foods sharing similar characteristics to be searched. For example, identification of all enriched, freeze-dried, or zinc-fortified foods, searches that are usually very laborious, can be easily performed using this method [11]. By unambiguously describing the foods in Handbook No. 8, this system ensures that the information retrieved from a food composition data base fits the users needs. This factored data forms the basis for current NCI initiatives developed to match foods and nutrient values among various data bases.

Table 1. Thirteen factors for the factored food vocabulary

1. *Product-type* – food group or category defined in terms of common consumption, functional, and/or manufacturing characteristics of a food
2. *Food source* – plant or animal from which food originates (e.g., chicken, wheat); for multi-ingredient food items, the food source of the main ingredient by weight (other than water) is indexed; other ingredients are included as additions under the factor term 'treatment applied'
3. *Part of plant or animal* – stem, flesh, root, organ meat, skeletal meat, etc.
4. *Physical state, shape, or form* – solid, liquid, divided into pieces; affects heat transfer through the product, susceptibility to permeation by chemical substances, and invasion by microorganisms; physical characteristics may be inherent or result from processing
5. *Degree of preparation* – amount of heat applied to a food product (e.g., raw, partially cooked, fully cooked)
6. *Cooking method* – process by which the heat or microwaves are applied to a food product (e.g., roasted, baked, toasted)
7. *Treatment applied* – processing steps usually involving adding, substituting, or removing food components or modifying food components (e.g., fermentation)
8. *Preservation method* – primary method used to prevent microbial and enzymatic spoilage
9. *Packaging medium* – substance within container where food is packed (e.g., syrup in canned fruits, brine in canned vegetables)
10. *Container or wrapping* – main container material and form and the material for the liner, lids, and ends
11. *Food contact surface* – container materials in direct contact with food
12. *User group* – consumer groups for which product is produced and marketed; age group and special diet characteristics
13. *Adjunct characteristics of food* – assortment of characteristics that apply only to certain foods (e.g., cut and grade of beef, color of poultry, and degree of plant maturity)

Committee on Data for Science and Technology

The establishment of INFOODS and the various vocabulary systems described has contributed to the development of data bases with more accurate and timely data. Existing vocabulary systems provide a firm basis for further work in the area of standardized food descriptions. Therefore, in 1986, it was proposed that the Committee on Data for Science and Technology (CODATA) establish a Task Group in the Systematic Nomenclature for Foods in Numeric Databanks. This task group was approved at the 16th CODATA General Assembly, September 1988, and met for the first time in January 1989 at a conference sponsored by NCI. It endorsed the idea of establishing a working group to address the need for a coding system for foods, incorporating a faceted thesaurus designed to comply with a number of precepts. One precept is that the system should be utilized by the end-user with a minimum of difficulty, both for retrieval and for coding personal or regional data. The coding system should also be applicable to a large centralized databank.

Data Base Development for Specific Dietary Factors

A major problem in assessing nutrient intakes throughout the world is the limited analytical accuracy and data completeness of standard food composition tables. The accuracy of composition information depends also on the methods of collecting and handling the data. There are sufficiently accurate data on the occurrence of protein and fat in most foods, but the data are limited for other important nutrients such as some vitamins, minerals, and fiber.

Currently, the DCB at NCI is sponsoring research on the development and improvement of analytical methods for detection and quantification of dietary fiber components and carotenoids. Epidemiological studies have demonstrated that increased dietary intake of these components has been associated with decreased cancer incidence. Foods that provide significant sources of dietary fiber in the US diet are being selected and analyzed for total dietary fiber and the major fiber components. To ensure representativeness of foods in the US diet, statistically based sampling techniques, emphasizing food type, geographical regions, genetic variety, processing, and environmental conditions, will be used. A comprehensive dietary fiber data base will be established that will contain composition values for total dietary fiber, cellulose, lignin, soluble and insoluble fiber, hemicelluloses, and pectins. These data bases will provide a useful tool for estimating the dietary fiber intakes of individuals and population groups, furthering the understanding of the role of dietary fiber in cancer prevention and control.

A project has been completed to study the methodologies for the determination and quantification of different forms of vitamin A and carotenoids in selected foods. In July 1985, a contractual arrangement was established to provide reliable data for the retinoid/carotenoid content of foods consumed by the US population. The objectives of this research project were to develop and validate analytical methods to identify and quantify retinoids and carotenoids in food, determine a statistically based representation of foods that are important sources of these nutrients in the United States, and develop a reliable data base of the retinoid/carotenoid content of these foods. Availability of the food composition data will allow the intake of these nutrients to be more precisely calculated and correlated with cancer incidence and mortality.

Data Surveys

Assessing the role of diet in cancer or other diseases has been limited by the comparative lack of reliable information on dietary intake patterns and corresponding composition of foods. Although food consumption surveys have been conducted in many countries, the data are in various locations, languages, and formats. Thus, comprehensive information regarding intake of nutrients or food contaminants remains inaccessible. Therefore, the DCB at NCI is conducting a 3-year project to collect international food consumption data and develop a system for nutritionists, epidemiologists, medical researchers, and other health professionals to use in assessing the intake of nutritive and nonnutritive food constituents.

The overall objectives of this effort will be to identify and obtain food consumption data from various countries and regions in the world, summarize selected surveys, and assemble the data into a computerized, international food consumption data base. Existing and available individual food intake data collected over the past several decades on free living individuals on self-selected diets will be obtained by the DCB from all available sources worldwide. Such data has been generated by academic institutions, government agencies, nutrition foundations and associations, commercial establishments, and other groups within countries. Although food industry data are not frequently published and/or difficult to access, this data will be included whenever possible. Countries will be selected on the basis of the existence and availability of acceptable cancer incidence and mortality data as well as appropriate dietary data of satisfactory quality. Cooperation of international organizations, including WHO, the International Agency for Research on Cancer (IARC), and officials of other governments, will help to compile these statistics.

Criteria will be developed to screen the data, including utility and integrity, statistical limitations, collection methodology, food identification, and retrospective versus prospective level of detail. Screened dietary survey data will be incorporated into a computer-based analysis data base. Information describing each food consumption survey, including methodology, population surveyed, dates, and summaries, will be documented.

To estimate intake of natural food components, for example, dietary fat or micronutrients, food additives, contaminants, or byproducts of cooking, international food component data will also be integrated into the system. The source of nutrient information will be determined for each survey. Standard nutrient tables such as US Handbook No. 8 will be incorporated. Procedures for updating the nutrient tables will be included so that system users can incorporate updates. On completion of the project, a computerized data base will be available in a PC-based desktop format that can be accessed using English commands by scientists around the world. Potential users of this data base include research epidemiologists and nutritionists, particularly in the area of cancer prevention, and various government agencies responsible for policy decisions and public health.

Conclusion

Information on the nutritive and nonnutritive components of foods is essential for our understanding of the role of nutrition in human health. The quality, quantity, and accessibility of these data constitute an important resource for health professionals concerned with the food supply and its implications for the overall health of individuals and population groups throughout the world.

Food composition data bases are already improving the power of many nutrition and health research efforts. In the future, as the food supply grows and as the relationship between diet and disease is further elucidated, the number of food data bases as well as their significant role in research and public policy will increase. Initiatives by NCI and other organizations to standardize and document procedures for maintaining timely and reliable international data bases will allow more appropriate comparisons between food consumption patterns and disease risk throughout the world.

References

1 Sorenson A, Lee HKM, Gloninger MF: Need for a standardized nutrient data base in epidemiologic studies; in Rand WM, Windham CT, Wyse BW, et al (eds): Food Composition Data: A Users' Perspective. The United Nations University, 1987, pp 41–51.

2 Borra S: Considerations for implementing dietary guidelines in the retail food industry. Guidelines on diets and health: implications and strategies for implementation. Presentation to National Academy of Sciences, Food and Nutrition Board, August 4, 1988.

3 Butrum RR, Young VR: Development of a nutrient data system for international use: INFOODS (International Network of Food Data Systems). JNCI 1984;73: 1409–1413.

4 National Academy of Sciences, National Research Council, Food and Nutrition Board: Diet and health: implications for reducing chronic disease risk. Council on Life Sciences. Washington, National Academy Press, 1989.

5 Rand WM, Young VR: International network of food data systems (INFOODS): report of a small planning conference. Food Nutr Bull 1983;5:15–23.

6 INFOODS final report for NCI contract NO1-CN-45168. July 1, 1984–October 31,1988.

7 Butrum RR: Considerations in designing a system for description of food and flexible retrieval of relevant data; in Glaeser PS (ed): Role of Data in Scientific Progress. CODATA. Amsterdam, North-Holland/Elsevier Science Publishers, 1987, pp 404–407.

8 Polacchi W: Standardized food terminology: an essential element for preparing and using food consumption data on an international basis. Food Nutr Bull 1986;8: 66–68.

9 Butrum RR, Pennington J: Technology systems used for food composition data bases: a historical perspective; in Glaeser PS (ed): Computer Handling and Dissemination of Data. CODATA. Amsterdam, North-Holland/Elsevier Science Publishers, 1987, pp 404–407.

10 McCann A, Pennington JAT, Smith EC, et al: FDA's factored food vocabulary for food product description. J Am Diet Assoc 1988;88:336–341.

11 Herold PM, Butrum RR, Brooks EM, et al: The National Cancer Institute food component research data base; in Greens R (ed): Proceedings of the Symposium on Computer Applications in Medical Care. New York, IEEE Institute of Medical Engineering, 1988, pp 512–517.

Ritva R. Butrum, PhD, Formerly Chief, Diet and Cancer Branch, NCI, NIH, 4082 Norbeck Square Drive, Rockville, MD 20853 (USA)

Simopoulos AP, Butrum RR (eds): International Food Data Bases and Information Exchange. World Rev Nutr Diet. Basel, Karger, 1992, vol 68, pp 15–26

International Food Data Base: Conceptual Design

Sudhir Srivastava, Ritva Butrum

Diet and Cancer Branch, National Cancer Institute, National Institute of Health, Bethesda, Md., USA

Contents

Introduction

Over the years, diet has been identified as a major factor affecting human health. Increasing evidence suggests that diet is closely associated with certain forms of cardiovascular disease and cancer, most notably breast, gastric, and colorectal cancers [1–4]. Epidemiologists and other investigators are becoming increasingly interested in researching the potential effects of diet on human health. In any quantitative study of human nutrition, information on the nutrient composition of foods is essential. It is required to assess the nutritional adequacy of diets, to calculate dietary intakes, to formulate dietary modifications, and to evaluate the role of dietary factors in the disease being studied. Further, it is often necessary to exchange nutrient composition data, both nationally and internationally, to obtain a more complete data base. This requires that the names and descriptive terms for the foods must be easily understood and unambiguous.

Although a major vocabulary system has been defined for feeds developed by the International Network of Food Information Centers (INFIC) [5, 6] there is no standard for developing an international data base with systematic nomenclature for foods. Until now, each country has developed a system to meet the legislative and regulatory requirements of its own government. Therefore, complexities of designing and maintaining an international data base require the cooperation of all countries involved in such an effort, and the National Cancer Institute (NCI) has taken a lead in this regard. The need for such a data base has been expertly discussed in the paper by Butrum and Young [7].

Further examination of this problem reveals that religious and socioeconomic factors greatly influence the type of foods consumed in various countries. A great number of synonyms and homonyms exist for a particular food as well as differences in the way people regard foods. The local name may be misleading to those not closely acquainted with the local language, culture, or legal system. Adding to the confusion is the introduction of processed foods that have changed Western society's food habits. The brand name of a particular food in one country may not be the same in another country. Even taxonomic nomenclature may be different from that used in other countries.

For example, the existing food tables produced in the United Kingdom and United States [8–10] may have limited use in other countries because the tables are more or less designed to meet the standards of those two countries. However, because of the increased volume of foods being imported from other countries, the growing number of international research collaborations, and the burgeoning number of ethnic foods now available, the United States and several other countries have recognized the necessity of having an international food composition data base that can be shared, stored, and accessed.

Research on the chemical composition of foods has resulted in a volume of data collected over the past hundred years. This information has resulted in many publications and scientific food tables [10]. However, the lack of a food description language and differences in the format of reporting units of measurement, and methods of analysis, have seriously limited the use of these food tables for comparing nutrient contents of particular foods [11]. Thus, no meaningful inference can be drawn from these tables in the international context, which provides us with ample reason to explore the design and implementation of an International Food Composition Data Base. This paper discusses the conceptual design of the International Food Data Base, introduces the software and hardware issues, and describes future cooperation among various countries.

International Food Data Base

The design of the International Food Data Base is an evolving task. It will require a close collaboration of experts from all over the world in evaluating and recommending methods for developing such an international data base system. The NCI has been actively involved in such an effort. Two international workshops of experts from various countries have recently been organized to advise and make recommendations to NCI regarding the implementation of the International Food Data Base [12, 13]. The outcome of these workshops has been a growing understanding of the problems that must be addressed before an international food data base can be realized. Although the idea of putting all data together into a single data base with access to everyone sounds appealing, there are several practical problems:

(1) *Copyright infringement:* Data bases created in various countries are copyrighted, which creates problems in reproduction for unrestricted use. One is limited in its use and inferences drawn from data. This also places administrative restriction on the access of data.

(2) *Technical problems:* There are several technical problems. For example, when and how will the data base be updated? What format of the data will be used to facilitate conversion?

(3) *Description of foods:* There are many coding systems and classification terminologies used in the world. Many of them are well adapted for the specific needs of the producing countries. Differences in foods and food cultures among countries, and even within a country, and differences in use for the data, have produced a multitude of coding systems. Consequently, an international coding system needs to be developed that would focus on utilizing and supplementing existing systems rather than replacing them. It should be noted, however, that it might be possible to convince food table developers and publishers to add coding of an 'international' unified system to their tables and data bases.

One of the major problems with the existing tables is the lack of information concerning the methodology used to assess nutrient content of foods. It is not always clear whether the reported values are from published literature or from the actual laboratory measurements collected by the reporting country. Also, there are no standard methodologies for measuring nutrients in foods. Which technique is best suited for a particular analysis? How should the results of the analysis be expressed? These are major questions that must be addressed before one can compare food tables from one country to another.

Therefore, problems are many, and the task of setting up an international data base appears overly ambitious. The two NCI international workshops [12, 13] have addressed many of these problems. Experts responsible for the management of food tables in their own countries reached a consensus on many issues and left unresolved issues for further discussion. It was suggested that despite some limitations to the food description language, LanguaL, it should be adopted for use in the International Food Data Base [13].

Food Description Language and Data Base Design

Food Description Language

The lack of an international food description language has always been a major impediment in realizing the International Food Data Base. In a recent meeting of NCI's International Food Data Base Workshop, the participants were of the opinion that the International Food Data Base should proceed with the use of LanguaL, formerly known as Factor Food Vocabulary (FFV). This language has been developed by the Center For Food Safety and Applied Nutrition (CFSAN) of the US Food and Drug Administration (FDA) together with NCI. A full description of the system can be found elsewhere [14, 15] and in the accompanying paper by Hendricks in this volume [pp. 94–103]. In brief, food characteristics are described by factors. Each factor has a set of related terms. Every factor term is defined in a dictionary of terms. By choosing one or more of these terms, a food's characteristics are precisely defined. Foods sharing similar characteristics contain the same factor terms in their LanguaL description. Each food described using LanguaL has, in addition to its descriptor term, a unique identifying number and name.

Presently, one of the drawbacks of the LanguaL system is that only one term is allowed for certain facets. This obviously limits the degree of detail with which a food can be described. However, the French [see paper by Max Feinberg, pp. 49–93] have successfully modified LanguaL in their data base to be able to incorporate more than one term per facet. The other limitation of LanguaL is its limited description of mixtures or recipes. In an effort to rectify this issue, NCI is working with FDA to begin full-ingredient coding of foods and to expand the language to allow more terms.

Data Base Model

Participating members of the 1990 International Food Data Base Workshop [13] agreed that a relational model will best serve the international community. This type of data base is increasingly popular because the relational model is simple to use with nonprocedural requests and provides data independence – because there is no positional dependency between relations, requests do not have to reflect any preferred structure.

The data base management system using a relational model was first introduced by Codd [16]. The relational model is based on relational algebra, that is, relation is a set with tuples as elements, and by definition a set is not supposed to have any duplicate elements. The details of the relational algebra are beyond the scope of this article. In simple terms, data are represented in a two-dimensional table, which is called a relational model of the data. In relational model terminology, a table is called a relation. Every column in a relation is an attribute. The row of the table is called a tuple. We could roughly say that the columns of the table represent data elements and the rows of the table represent the data records in conventional terminology. A column or set of columns is called a 'candidate key' or just a 'key' when its values uniquely identify the rows of the table. It is possible for a relation to have more than one key, in which case it is customary to designate one as the 'primary key'. Thus, in a table with uniquely assigned names for the columns, the ordering of columns and the ordering of rows are insignificant. The ordering of a table is insignificant, too. The tables are produced after careful observation of the entities to be included in the table. The process of crystalizing the entities and their relationship in table formats using relational concepts is called the 'normalization process'. The normalization theory is based on the observation that a certain set of relations has better properties in an updating environment than do other sets of relations containing the same data.

Software

Choosing a Data Base Management System

As mentioned earlier, the approach to designing this data base will be based on a relational model. The Relational Data Base Management System (RDBMS) has emerged as a reliable, flexible, and efficient means of managing data. Unlike other data models, such as hierarchical and network models, in which each element of the data sets is associated with pointers, thereby imposing a fixed relationship, the relational data base model is free from any such restriction and is capable of handling different types of queries. Because this relational data base model treats all information uniformly as data elements, stored as rows of tables, any value can be used to associate or join one table with another. Thus, a relationship between values is dynamically created. Some of the salient features of RDBMS are:

(1) Flexibility in data modeling.

(2) Consistency in data by reducing duplication.

(3) Independence of physical data storage from logical design.

(4) A high-level language for data definition, manipulation, and query Structured Query Language (SQL).

(5) An open architecture facilitating systematic data sharing and ease of developing applications.

There are several software vendors supporting relational data base management systems, such as Oracle, InfoDB+, DBASE III+, Unify, Ingres, and DB2. The selection of software was greatly influenced by cost and by versatility of its use. Also, the selection of software was based on the vendor's ability to provide:

(1) Techniques for using software packages.

(2) Ways to think ahead for expansion purposes.

(3) Hardware requirements.

(4) Documentation for RDBMS.

(5) Program service support and maintenance.

(6) Potential application areas.

(7) Implications of the vendor's future plans on the current or planned software/hardware.

(8) Educational training.

(9) Hardware configuration supported.

(10) Ancillary packages, for example, data dictionary, aids, applications, data communication interfaces, and monitors.

Oracle has been chosen to implement the data base. Selection of this software in no way reflects its superiority over other RDBMS software. However, it meets specific goals and was accepted by the international collaborators. Some of the features that Oracle provides meet future expansion plans for the data base server and network approach. Oracle provides RDBMS's capable of providing all the facilities one expects of a true RDBMS such as the joining or projection of one table with another. Oracle also provides the following:

(1) SQL*Form for form-based transaction processing form and SQL*Plus and SQL*Report for ensuring data integrity.

(2) Several Case Tools and 4GL, which can be used to analyze the impact of any changes and generate functional prototypes and production applications.

(3) An active data dictionary that easily can be modified to meet the changing needs of an evolving data base.

(4) The capability to interface with several 3D languages such as COBOL, C, Fortran, etc.

(5) Tools that can be used to create prototype and production applications, provided by several third-party vendors.

Because we expect to have access to this data base by the users via linkage to the in-house computer, Oracle has a significant advantage over other RDBMS's. Oracle can support distributed processing and data base by using SQL*Net, SQL*Connect, and Oracle Servers:

(1) SQL*Net is a remote procedure call (RPC) program that resides on both client and server machine. It provides the necessary network communication links between the application and the data base server and among multiple Oracle servers via the network transport level.

(2) SQL*Connect is a 'data base gateway' that lets Oracle systems talk to other vendor's servers. It functions primarily as a translator of SQL requests and error codes.

Conceptual Model of the Data Base

Based on our preliminary data analysis, we have constructed a conceptual model of the International Food Data Base that shows the data entities and the relationship between them. There will be eight major tables, excluding minor tables, that may be needed to establish relationships between tables through keys. The conceptual overview and interrelation

Table 1. Data record for international food data base[1] (descriptions are provided for those not clear)

Table name	Variable and length	Description
FOODS	FID#, char, 5	Food identification key, no null value
	ORGID#, char, 5	Origin identification key, no null value
	Name, char, 100	Name of the food
	Geo-area-char, 4	Place where food grown or produced
	Org-Part, char, 100	Part of plant or animal
	Part-Cond, char, 100	Physical state, shape, or form
	Dietary-Use, char, 100	
	Processing, char, 100	Type of food processing
	Refuse Value, char, 100	
	Text-Definition, char, 100	
SYNONYMS-FOODS	FID#, char, 5	As defined above
	Syn-FID, char, 20	Synonym for the food identification number
	SID#, char, 5	Synonym identification number key, no null value
COMMON-NAME	FID#, Common-name, char, 20	As defined above
ORGANISMS	ORGID#, char, 5	As defined above
	Family, char, 30	Name of family
	Genus, char 30	Name of genus
	Species, char, 30	
	Variety, char, 50	
SYNONYMS-ORG	ORGID#	As defined above
	Syn-Genus, char, 30	
	Syn-Species, char, 30	
	Syn-variety, char, 30	
MEASURES	FID#	As defined above
	CID#, char, 5	Component identification key, no null value
	Average, real, 5	
	Std-Dev, real, 5	Standard deviation
	Samp#, int, 3	
	Method, char, 50	Method of analysis
	SID#	As defined above
COMPONENTS	CID#, Name, char, 50	As defined above
REFERENCES	SID#	As defined above
	Title, char, 150	Title of the paper
	Author, char, 100	Name of author(s)
	Year, int, 4	Year of publication

[1] Variable types are not necessarily in the Oracle format. For example, variable type number (in Oracle) is for both integer and real data types. Also, character variables of unknown length will be accommodated by using *var char* type in Oracle.

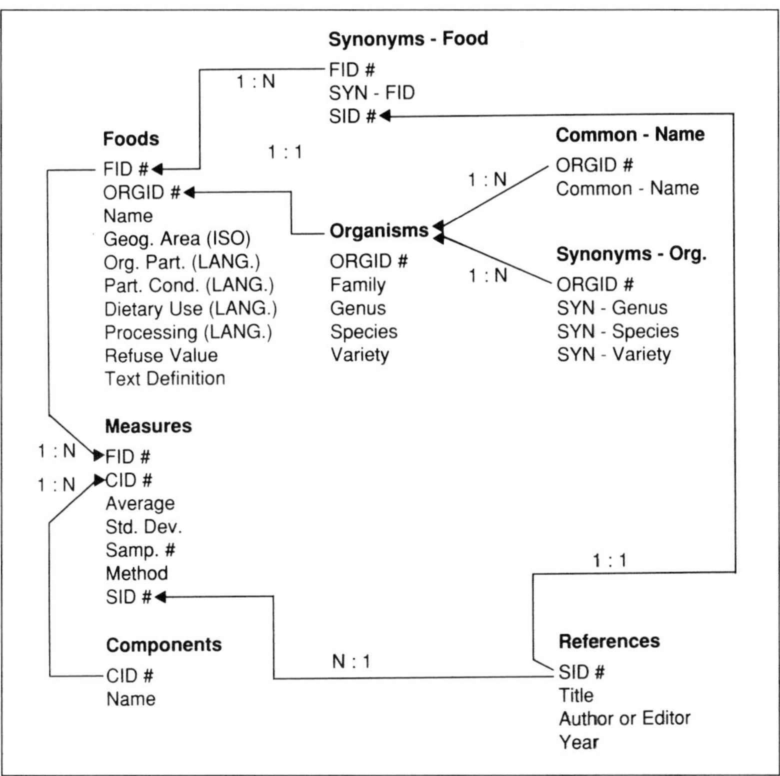

Fig. 1. Representation of the conceptual view of the International Food Data Base. Name in bold letters represents the name of a relation or a table. Keys in each relation or table are denoted by the pound sign (#). Relations between attributes are shown by arrows for either one-to-one (1:1) or one-to-many (1:N or N:1) relationship.

between tables are shown in figure 1. The major components (attributes) of each table are also shown in this figure. The file structure of these tables is shown in table 1. The major tables are FOODS, SYNONYMS-FOODS, COMMON-NAME, ORGANISMS, SYNONYMS-ORG, MEASURES, COMPONENTS, and REFERENCES. As we progress, several ancillary tables may be added providing information on the processing, packaging, illustrations, and ingredients of the food. The information stored in these tables will be from published literature.

The relationship between entities is part of the conceptual model, and it must be represented in the data base. Any number of entities can participate in a relationship. There also exists mapping between attributes of an entity. The mapping of such a relationship is shown in figure 1. There are mainly two types of relationship in the present data base:

(1) One-to-One: Such a relationship exits between SID# (SYNONYMS-FOOD) and SID# (REFERENCES). This is denoted by 1:1 on the top of the arrow.

(2) One-to-Many: Such a relationship exists between many of the attributes. They are denoted either 1:N or N:1, depending on the direction of the attribute with many-to-one relationship.

The name of the attributes is shown for each table. Wherever applicable, we also have shown the methods of representing their values. For instance, we will use three-character ISO (International Standards Organization) acronyms for geographical regions and LanguaL for Org-Part, Part-Cond, Dietary-Use, and Processing. A major benefit of this data base system is that many of the elements including LanguaL have been developed. Additionally, a thesaurus of plant names and synonyms has been developed in five languages by Haendler [17].

Users of Data Base

The proposed data base is intended for a wide spectrum of users. Some of the potential users of this data base may include investigators conducting dietary intake and consumption studies, clinical trials, and epidemiological studies. The data base may thus provide useful information for dietary intake and food consumption studies. Similarly, in clinical trials, particularly in those studying ethnic groups or trials conducted in a foreign country, the data base may provide valuable information and data needed to assess dietary intake and compliance during trials. This stand-alone system may later be tied into several nationally known data bases, such as Natural Products Alert (NAPALERT), USDA GERMPLASM Data Base, and other data bases containing phytochemicals relevant to cancer prevention.

Conclusions

Presently, the International Food Data Base is in the developmental phase. However, since many of the system's components are now in place, and since FDA has initiated full ingredient coding of foods, it is possible that this system could be developed with minimal time and effort. The time is right to develop such a system as the national and international data base communities recognize the need for this system and have expressed great interest in working together. At this time, the countries actively participating include France, the United Kingdom, Denmark, Hungary, Poland, Italy, Guatemala, Japan, Greece, Taiwan, China and India. Our approach will be guided by the experience of these international collaborators, and modifications will be introduced if necessary.

Despite some shortcomings, the development of the International Food Data Base can proceed parallel to the resolution of the problems. It will be ensured that the developed data base provides flexibility for incorporating future changes and serve as a 'template' for the data exchange from various countries.

References

1 Pyorala K: Dietary cholesterol in relation to plasma cholesterol and coronary heart disease. Am J Clin Nutr 1987;45:1176–1184.

2 Palmer S: Diet, nutrition and cancer: Progress in food and nutrition. Science 1985;9: 283–341.

3 Jones DY, Schatzkin SB, Green, F: Dietary fat and breast cancer in the National Health and Nutrition Examination Survey. I: Epidemiological Follow-Up Study. JNCI 1987;79:473–485.

4 National Research Council: Diet and Health. Washington, National Academy Press, 1989.

5 Harris LE, Haendler H, Riviere R, et al: International feed databank system: An introduction into the system with instructions for describing feeds and recording data. International Network of Feed Information Centers. Publ No 2, 1980. Prepared on behalf of INFIC by the International Feedstuffs Institute, Utah Agricultural Experiment Station, Utah State University, Logan.

6 Haendler H, Echterdinjen L: Synthetic description systems for accurate data identification and selection. Principles and methods of nutritional data banks. Int Classif 1988;15:64–68.

7 Butrum RR, Young VR: Development of a nutrient data system for international use: INFOODS. JNCI 1984;73:1409–1413.

8 Paul AA, Southgate DAT (eds): McCance and Widdowsons – The Composition of Foods. London, HMSO, 1978.

9 Paul AA, Russell EC (eds): McCance and Widdowsons – The Composition of Foods (suppl 1). New York, Elsevier, 1981.

10 Heintze D, Klensin JC, Rand WM (eds): International Directory of Food Composition Tables, International Network of Food Data Systems, United Nations University, 1988.

11 Butrum RR, Pennington JAT: Technology systems used for food composition data bases: a historical perspective. 10th Int CODATA Conf, Ottawa, Canada 1986; in Glaeser PS (ed): Computer Handling and Dissemination of Data. Amsterdam, North-Holland/Elsevier, 1987, pp 404–407.
12 NCI International Food Component Data Base and CODATA Task Group on a Systematic Nomenclature for Foods in Numeric Databanks, Jan 30–Jan 31, 1989, held at the National Cancer Institute, National Institutes of Health, Bethesda, Md.
13 Fourth Annual International Data Base Workshop, March 28–March 30, 1990, held at the National Cancer Institute, National Institutes of Health, Bethesda, Md.
14 US Food and Drug Administration: Langual user's manual. Center for Food Safety and Applied Nutrition, Division of Information Resources Management, Information Support Branch, 1989.
15 McCann A, Pennington JAT, Smith EC, et al: FDA's factored food vocabulary for food product description. J Am Diet Assoc 1988;3:336–341.
16 Codd E: A relational model for large shared databanks. Communication of the ACM 1970;13:6–10.
17 Haendler H: Konzipiert für die Belange der Datendokumentation: der Internationale Futtermittelthesaurus; in Studien zur Klassifikation 1985;14:167–174.

Sudhir Srivastava, PhD, Diet and Cancer Branch, National Cancer Institute, National Institutes of Health, Bethesda, MD 20892 (USA)

Simopoulos AP, Butrum RR (eds): International Food Data Bases and Information Exchange. World Rev Nutr Diet. Basel, Karger, 1992, vol 68, pp 27–48

Principles for the Preparation of Nutritional Data Bases and Food Composition Tables

D.A.T. Southgate[a], *H. Greenfield*[b]

[a] AFRC Institute of Food Research, Norwich, UK;
[b] University of New South Wales, Sydney, N.S.W., Australia

Contents

Introduction

In general, the principles involved in the preparation of nutritional data bases and food composition tables are virtually identical, and in this chapter the term 'data base' will apply equally to computerized data bases and printed food composition tables. Where the two require different approaches, data bases will be qualified as either computerized or printed, as appropriate [1].

It is extremely rare for someone (or a group) preparing a compilation of nutritional data to start completely de novo, as the early workers in the field did, for in most cases there is an existing compilation, usually in a printed form, that is the starting point for his or her work. This may not apply to the actual structure or format of the data base where the task often includes the objective of improving or expanding the existing data base.

In the present discussion of the process of preparing a data base the existing data base would be evaluated in the same way as all other sources of published data. This means that the principles followed in revising and updating a data base are indistinguishable from starting with a clean sheet of paper or computer file. The only difference lies in the volume of work required to acquire data.

The work of preparing a data base can be considered to fall into a series of stages [2]: (1) Evaluation of the requirements of the potential users of the data base. (2) Development of a list of the range of foods that will be included in the data base. (3) Identification of the range of nutrients and other constituents whose composition will be given. (4) Acquisition of the compositional data. (5) Evaluation of the data. (6) Compilation of the data base. (7) Validation of the data base against the users' requirements.

Although these stages clearly form a sequence of tasks, the actual preparation involves interaction between the various stages. For example, the selection of food items will need to be judged against the users' needs and the evaluation of the data may require further analytical or literature work in acquiring data.

Identification of the Requirements of Users

One important principle is that data bases are essentially 'tools', and nutritional data bases are essential for a wide range of nutritional tasks, from the management of metabolic disease states to nutritional research and teaching. They are especially important in the epidemiological study of the relation between diet and disease [1, 3]. The data bases are used by governments for the assessment of the adequacy of the food supply and dietary intake of populations and increasingly in the development of food and nutrition policies. Nutritional data bases in many developed countries also form the basis for the nutritional labeling of foods and food products.

These users require the data base to meet their specific requirements, and it is therefore essential in the early stages of the preparation process to have a very clear idea of the types of uses of the data base. The specifications for the data base and the structure and format must be decided on the basis of the users' needs.

The range of foods required and particularly the level in the food chain at which consumption is measured will clearly be defined by the types of use. Governments and international agencies often have to make nutritional judgments and estimates of the adequacy of the diet using information on the amounts of food measured at the wholesale commodity level; whole cereal grains measured in tons, meat recorded in terms of carcasses at specified grades or primal joints, and many primary food commodities that form ingredients in food manufacture [4]. This type of use requires a limited number of food items with compositional data corresponding to the raw commodity.

The use of the data base for nutritional labeling necessitates including composition of foods 'as purchased', including raw foods, precooked foods, and branded products. Data at this level are also widely used for nutritional studies where household purchases are measured to provide estimates of food and nutrient intakes [5].

Measurements of the intakes of individuals and the prescription of diets for the treatment or management of disease require foods whose composition is recorded 'as consumed', and this makes considerable additional demands on the preparers of data bases who consequently have to consider the inclusion of cooked dishes that are complex mixtures prepared according to a range of differing recipes.

The different users also have differing requirements for the coverage of nutrients and other constituents, and these determine the volume of compositional measurements required and also have implications for the structure and formats of data bases. The compositional information required for measurements at the raw commodity level is usually very limited; indeed, the inclusion of very detailed nutrient compositions at this level could be highly misleading because of the inevitable losses of food and nutrients during progress through the food chain. At the 'as purchased' level, the range of nutrients that many users require is very extensive, even though the estimation of the intake of labile nutrients at this level of consumption is at best approximate. With the 'as consumed' food items, the coverage of nutrients required is usually essentially complete, although again, the values for labile nutrients and some trace constituents, whose concentrations in foods are extremely variable, can only be regarded as approximate indicators.

The use of data bases in a quasi-statutory role in connection with nutritional labeling introduces other user requirements that need to be included. Because all foods are biological materials, they show natural variations in composition, and if a published value is to be used in food labeling, ideally, the data base should include some estimate of variance so that if a value is challenged it can be assessed whether the value lies within the expected analytical range.

There is growing recognition that many foods, especially plant foods, contain minor constituents that are biologically active and may interact with nutrients to alter the bioavailability of a nutrient or may have biological effects that relate to health. Therefore, the intake of these minor constituents is potentially of interest in the study of diet and disease. There is a need to expand the coverage in nutritional data bases to include the needs of researchers in this area [6].

Analogous arguments have been used to support the inclusion of contaminants and food additives in nutritional data bases and in many fields of research, for example, studies of food intolerance. The only objection is that the concentrations of these constituents, especially contaminants, are variable and the distributions in foods are skewed so that mean or even median values are of little value in assessing intakes. The concentrations of additives are very much in the control of the food producer and show variations due to technological requirements. Some manufacturers believe that commercial advantages would be lost if additive levels in their brands were identified. However, unless brands are identified in the data base, the value of including the data would be lost, because the inclusion of average values for additives would be of little value to those who wish to use such data.

The uses to which the data base will be put also specify the levels of accuracy and precision that are necessary for the specific uses. Some uses require only moderately accurate, semiquantitative information whereas others require the maximum possible accuracy. However, the natural variability of foods places limitations on the predictive accuracy that can be achieved using data bases, and it is important for users of the data base to have a realistic view of what it can provide. In some cases, a data base cannot be expected to achieve the level of accuracy required, for example, in metabolic work or studies of labile vitamins and many trace elements. The statutory usage of data bases in many instances is inappropriate where regulatory decisions are to be based on values given in a data base using absolute standards of correctness.

Finally, the data base preparers need to have a sound appreciation of how it will be used because this will influence format in a published data base or the type of data base management system that would be appropriate in a computerized data base. If, for example, the data are going to be interrogated to find the composition of foods, then one type of data base format would be adequate; on the other hand, if the data base is to be used for a wide range of types of calculation, then another system may be required.

Establishing a List of the Food Items to Be Included

In all countries of the world the number of food items that are consumed is very large, especially if their composition is to be recorded as the raw basic commodities, 'as purchased' and 'as consumed'. In the case of cooked composite dishes, the range of recipes that might be used and are candidates for inclusion, adds to the number of food items. As an estimate based on the numbers of food items in a typical retail supermarket, complete coverage for a country such as the United Kingdom would require numbers of the order of 100,000.

The logistics of preparing a data base of this size mean that the compilers have to establish priorities for the food items to include. As a general principle this should be based on food consumption statistics so that those foods that are consumed in the greatest amounts by most of the population for whom the data base is being constructed have the highest priority. The next group would be those foods forming the major part of the diets of groups within the population, including infant formulas and weaning foods and specific foods restricted to ethnic minorities.

The third group of priorities would include foods that are major sources of specific nutrients in the diet but that are eaten infrequently. Finally, the data base should include as many examples as possible, within the resources available, of foods that are not consumed in sufficient quantity to be considered first priority (fig. 1).

Where a data base is being developed de novo, the principle of first including examples of the major food categories and then subsequently expanding the foods within the group is one way of developing a data base that can be used in a limited way early in its preparation.

As an example of this, one could imagine a hypothetical data base containing the items meat, fish, bread, flour, fruits, vegetables, milk, cheese, fats and oils, eggs, and sugar. In each case the food item is a mixture based on frequency of consumption. Such a data base would give appropriate estimates of the energy, protein, and fat intakes of the population as a whole. An even simpler data base could contain just two items, foods and beverages, where each item was a composite of the foods and beverages mixed according to consumption. This data base would give an estimate of the population's intake that was as good as the estimates of average food consumption. Although these are extreme examples, they illustrate the principle that the first aim should be for complete coverage at a broad level and then develop the detail, and second, the results from the use of a data base will only be as good as the estimates of food intakes.

The users' views on the proposed list of foods need to be considered, and it is important to establish very clearly the priority that will be given to each food item as objectively as one can.

Food Identification

Although it is not essential at the start of the preparation program to decide on how foods are to be grouped or identified in the final data base, it is useful to develop a strategy early. The correct identification of the food items and their relation to the compositional data are key parts in the correct usage of the data base, and unequivocal assignation of data to a food item is essential for proper data interchange between data bases. The Eurofood group [7] has developed a system of food coding, Eurocode, which attempts to address this issue, and in the United States the factored food system, which has its origins in the INFIC system of animal feedstuffs data base, is seen as making food identification accurately [8]. All systems of identification, either by coding or nomenclature, need to be carefully evaluated. Coding had its origins when the use of an alphanumerical system placed unacceptable demands on computer memory and, in principle, identification of foods by their name alone would not test most present computer systems. The major difficulty is actually agreeing on the name,

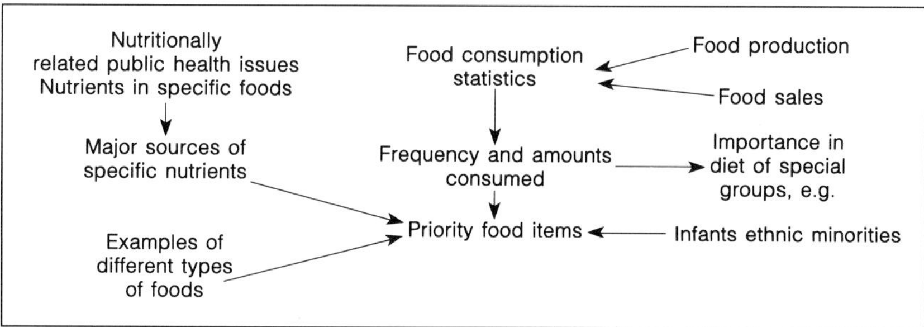

Fig. 1. Flow diagram for establishing priorities for food items.

even within one culture and a relatively small country such as the United Kingdom, when the same food is known by different names and different foods are known by the same name. The problem of nomenclature is magnified when the food is named in a different language because the best literal translation can give the same name to quite different foods. Some kind of systematic naming analogous to the systematic names of species is probably the best solution, and the factored food system is a polynomial system in essence as is Eurocode, which uses numbers instead of names.

In a statutory data base within one country, the bar code system could be used to identify specific foods, but this system is limited to products with bar codes and could not be used for all foods nor in all developing countries. The identification of commodities can be based on systems developed for the fiscal control of exports and imports.

Identifying the Range of Nutrients and Other Constituents to Be Included

To satisfy the wide interests of the users usually it will be necessary to include all the nutrients for which recommended intakes have been developed. This means that full coverage of inorganic nutrients and vitamins will be necessary. The issue of establishing priorities is less important for nutrients than it is for foods because, although the number of potential food items is very large, the number of nutrients is reasonably finite, although this is not the case when nonnutrient components are to be included.

Most nutritional users, particularly the research epidemiologists and many clinical nutritionists, will require fatty acid compositions, cholesterol, and characterization of the carbohydrates into at least sugars, starch, and nonstarch polysaccharide (dietary fiber). For carbohydrates, the preferred methods of analysis generate data on the species of sugars and the composition of the nonstarch polysaccharide so that the logistical demands of the analysis are the same for the more detailed coverage. Clinical nutritionists working in the renal field require amino acid compositional values, and these are also of value in assessing protein quality in countries where protein intake is limited. Table 1 illustrates the range of nutrients that would cover most types of use [1].

In considering which other constituents to include, the list is virtually unlimited. Possibly the first priority should be given to those substances that are known to influence the bioavailability of nutrients, such as phy-

Table 1. Nutrients and other constituents to meet most data base users' needs

High priority for most users	Desirable for nutrition research	Desirable for future developments
Water		
Total nitrogen		
Protein (total N – nonprotein derived N × factor)	Protein – N Nonprotein – N (NPN)	Components of NPN
Amino acid composition		
Fat, total	Fat, glyceride Phospholipids Sterols	
Fatty acid composition	Isomers of unsaturated fatty acids	
Carbohydrate, total		
Sugars, total	Individual, mono, di-, tri-, tetrasaccharides Sugar alcohols	
Starch		
Dietary fiber, total	Noncellulosic polysaccharides (NCP) Cellulosic Lignin	Monosaccharides composition
Organic acids, total	Individual organic acids	

Table 1 (continued)

High priority for most users	Desirable for nutrition research	Desirable for future developments
Alcohol		
Energy[1]		Measured values of gross energy content
Sodium		
Potassium		
Calcium		
Magnesium		
Iron		
Zinc		
Copper		
Phosphorus		
Chloride		Nitrate, nitrite sulfate, essential
	Iodine, fluorine	trace elements: Cr, Mn, Se, Co
		Contaminants: Pb, Cd, As
Ash		Hg, Ni, Al
Vitamin A, retinol	Retinoids	
Carotenes	Isomeric forms	
Total vitamin A activity		
Vitamin D	Values for D_2 and D_3	
Vitamin E, tocopherols	Isomeric forms present	
Vitamin K		
Thiamin		
Riboflavin		
Niacin		Free and bound forms
Niacin activity, total		
Folates, total		Separation of isomeric forms and glutamate conjugates
Vitamin C	Ascorbic acid Dehydroascorbic	
Vitamin B_6	Pyridoxol, Pyridoxal, pyridoxamine	
Pantothenic acid		
Biotin		
		Organic contaminants
		Natural toxicants
		Additives
		Components with physiological or pharmacological properties

[1] These values are derived by calculation.

tates, tannins, and oxalates, and then the constituents that possess physiological properties so that their consumption would modify the responses to diet. This would include the nonprovitamin carotenoids, plant sterols, saponins, glycoalkaloids, and glucosinolates.

At this stage it is useful to consider the units to be used in the data base while these decisions can still be delayed; if data are initially collected in a range of units the possibility of later confusion and errors is increased. The INFOODS manual on nomenclature for nutrients and other constituents [9] provides guidance on the preferred, usually internationally agreed, nomenclature for nutrients and the preferred units for expression. The adoption of these guidelines will greatly facilitate data interchange between data bases and lead to compatibility between data bases, which is essential for international studies of nutritional epidemiology and also allows the nutritional community to maximize the use of the resources available for nutritional analysis of foods.

The Acquisition of the Compositional Data

The first three stages of the preparation process define the specifications for what the data base will contain in terms of food items and constituents. The fourth stage is concerned with getting the data to fulfill this specific (table 2).

When starting a data base de novo one option, an indirect option, is to decide that the data in some other data base is adequate for your use and merely use this data in the format most appropriate to your needs. This would be a reasonable option, only if the foods grown, produced, prepared, and eaten in the other country or region were identical to your own; clearly such a situation would be very rare and this course of action would be acceptable if resources were so restricted that the specific requirements of the local users had to be ignored.

The other options were characterized by Southgate [3] as direct: where the foods for the data required are sampled and analyzed specifically for the construction of the data base. This is the best option if resources are available because the data base compilers have control over the identity of the foods sampled to ensure that the sampling is representative and offer the choice and execution of the analytical operations to ensure the quality of the data. In the indirect option the compilers use data that have been obtained for other purposes or that have been published in the scientific

Table 2. Specifications of the data base

Major specification	Subsidiary specification	Data base coverage
Coverage of food items	Level of compositional measurement Food nomenclature	Wholesale As purchased As consumed
Coverage of constituents	Nutrients	Proximates Minerals Vitamins
	Minor biologically active components	Those that interact with nutrients
		Physiologically active components Contaminants Additives
Modes of expression	Units Conventions for vitamins	Conform with statutory requirements
Operating systems	Search facilities Calculation	Indexing Analysis

literature or other data bases and use unpublished data where their provenance is known. This option requires careful scrutiny of the data to establish that they are appropriate for the data base against a series of criteria (table 3) [1, 3].

The Direct (Analytical) Method of Obtaining Data

As mentioned above, this is the preferred method because of the control that it provides over the source, identity, and quality of the data. It is, however, very demanding of resources and should only be followed when a number of conditions have been satisfied. The first condition is when there is absolutely no suitable information on the composition of a food, which may be true for a novel processed food product or when dietary studies are being made in a country where such studies have not been made before.

A second condition is when there is evidence that the composition of a food has changed significantly since it was last analyzed. This evidence may be presumptive or based on knowledge that the production or processing of the food has changed in a way that is likely to have affected its

Table 3. Criteria for scrutiny of data

Parameter	Criteria
Sample identity	Unequivocal identification of sample
Sampling protocol	Collection of representative sample
Sample preparation	Methods used
	Precautions taken
	Material rejected as inedible, etc.
Analytical sample preparation	Nature of material analyzed
Analytical procedures	Choice of method
	Quality-assurance procedures
Mode of expression	Compatible with that used within the data system

composition. Decisions based on presumption are most justified where the food in question forms a major part of the diet or is growing in importance.

The last condition is when there is evidence that the analytical procedures used in the past were flawed or where users require information for constituents not previously measured and included in the data base [10].

The Indirect (Literature-Based) Method

The previous section includes the implicit argument that the direct method, because of the demands that it makes on resources, should only be used when there is either no data or when the existing data are inadequate. It therefore assumes that some information exists and in this respect merges with the indirect method based on the use of data from published and unpublished sources.

Strategies for Acquiring Data

It is possible to establish the preferred strategy by reference to the decision tree (fig. 2). The first stage is to assess existing data for the food in question; if none exist then sampling and analysis is the only option. If data exist, they are assessed using the criteria listed in table 3, which provide the compilers with indications of the quality and appropriateness of the data. If the data are judged inadequate, sampling and analysis are required.

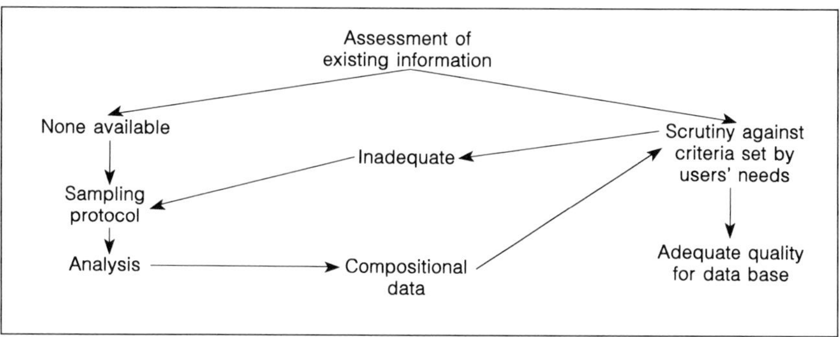

Fig. 2. Decision tree for strategies for acquiring data.

Sampling and Analyses

It is appropriate to consider these two operations together because correct sampling is essential if the analytical values are to be of any value [1]. Furthermore, the handling of the sample when it is collected and during transfer to the laboratory can have a profound influence on the composition of the food and the validity of the analyses. The logistics of the analytical operations also need to be closely integrated into the sampling protocols to ensure that the intervals between sampling and analysis are appropriate.

Sampling protocols should be designed to ensure that the samples collected are representative of the food at the level planned. For example, if values representative of the food eaten throughout the country are required, a countrywide sampling frame is necessary; if there are seasonal effects on the food, the sampling frame will need to consider these; where prepared or processed foods are being studied, the samples must be representative of the variations in production and preparation methods. Sampling protocols therefore need to be developed against an understanding of the foods in question and the patterns of their production, preparation, and consumption.

The sampling protocols should include details of how the samples are to be treated once they have been collected and at all the stages up to their analysis. These operations need to be designed in collaboration with the analysts; in fact, there are advantages in making the analysts responsible for the sampling protocols because they are familiar with the stability of nutrients and precautions essential to prevent contamination and changes in composition.

The analytical protocols themselves should define the methods to be used. The overriding principle in the choice of analytical methods is the requirement that the values obtained be nutritionally relevant. This will most often decide the principles of the methods used but not necessarily the precise method because for many nutrients there are a range of satisfactory procedures, giving essentially the same values and similar analytical characteristics such as reproducibility and repeatability. Where the values in the data base are intended to be applicable in nutritional labeling, there may be statutory regulations that define the methods to be used for the measurement of many nutrients. Clearly these requirements relieve the data base compilers of making decisions on analytical methods. The requirements for nutritional research, however, may make the use of other procedures or the development of specific procedures for nutrients or species of a nutrient that are not covered by these legal obligations desirable.

It is important that the methods used to obtain nutrient or other compositional values are documented fully within the data base to enable the users to know precisely what has been measured. Also, during the life of the data base, new methods may be used and subsequent analytical study may show that the values obtained by the different methods are not compatible. This documentation will identify the values that need to be verified during revisions.

In choosing the analytical methods to be used, the resources, both of personnel and material, may be deciding factors [1]. For example, although many modern analytical methods demand the use of extensive instrumentation and therefore capital resources, it should be remembered that many manual analytical procedures that need simple apparatus and are labor intensive can give acceptable data for many nutrients. Some nutrients cannot be measured satisfactorily without instrumentation, and when resources for the preparation of a data base are limited it may be necessary to accept that literature values for these nutrients will be used.

Greenfield and Southgate [1] have reviewed the range of available methods for the measurement of most of the nutrients given in table 1 and indicate the methods that appear most suitable for the production of data for nutritional purposes.

The major determinants of the validity of nutritional data are the principles that form the basis of the method, 'what the method sets out to measure', the validity of the sampling protocols, and the attention given to quality assurance in the analytical work.

Evaluation of the Data

The process of evaluating data was introduced earlier when discussing whether to embark on an analytical program or to rely on the indirect literature approach. This section discusses the evaluation process in more detail and gives some thoughts on the way the evaluation process should be carried out.

The evaluation process is undertaken to ensure that the data included in the data base are of the quality required by those who will use the data base. This is done by establishing and applying criteria that are as objective as possible, which determine whether the data are of sufficient quality for inclusion in the data base. Quality can be defined as fitness for the purposes for which the data will be used and thus depends on the users' requirements for the data base. For some nutritional purposes a semiquantitative series of data is adequate; for others a much higher level of accuracy of the values in the data base is seen as essential. This would be true for statutory nutritional labeling. In some respects a data base prepared for nutritional use diverges from a data base for nutritional labeling at this point because for virtually all nutritional uses the accuracy of the data base is not the major factor in limiting its accuracy for assessing nutrient intakes or when formulating diets.

Criteria for Evaluating Data

These criteria (table 3) are applicable to all data intended for inclusion in the data base, whether the data have been generated specifically for the data base or taken from the literature (including any existing data bases) or unpublished sources.

Identity of Food Items

This involves considering the sources of the samples and descriptions of the food that was analyzed. Such information is essential for ensuring that the food is correctly identified and is compatible with the food item that will be measured or used by the data base users.

Sampling Protocols

These show whether the sample was representative and add further to validating its identification. The sampling protocol also defines the precautions taken to eliminate contamination and changes in nutrient composition from collection to analysis.

Sample Preparation

This, taken with the first two criteria, further serves to identify the material analyzed and the proportion of inedible waste removed before analysis.

Analytical Procedures

The analytical procedures used define which substances have been measured and whether these meet the nutrient specifications set up in consultation with the users. For example, the users may require nonstarch polysaccharide values whereas the data being evaluated may be total dietary fiber values measured by the AOAC gravimetric method.

Analytical Quality Assurance

This defines the quality of the analytical results and shows whether the methods have been properly described and tested, preferably in collaborative trials. It is necessary to establish how the methods were controlled in use, for example, by the analysis of reference materials and standards. In high-quality analytical work it is possible to conduct an audit of all the stages followed in the generation of the data.

Application of the Evaluation Criteria

All data should be evaluated against the criteria to ensure that they meet the needs of the users and the specification of the data base. Where the data have been generated specifically for the data base, the application of the criteria is relatively straightforward because the documentation of the samples and the sampling and analytical protocols will have been set and it only remains to ensure that they have been followed. When literature data are being evaluated the process can be more difficult because the documentation of the foods analyzed and the descriptions of the sampling and analytical protocols may not be adequate. This is frequently the case with food consumption compilations and emphasizes the need to document the data as fully as possible during compilation. Ideally the provenance of all values should be established.

The evaluation of data is probably best carried out in a sequence of stages that involves iteration between them as the evaluation proceeds. This approach to evaluation ensures that judgments about the data are formed progressively and objectively. The stages are concerned with evaluating primary sources of data, with compiling an archive of all the existing data on a food item, and with assembling the data into a comprehensive data base and finally in the data base(s) for the users. At each stage, the criteria are used in a different context that imposes additional criteria against which the data can be assessed.

Evaluation of Data Sources

This stage involves applying the criteria to the data in the data source, which could be an analytical report, a published paper, or an existing data base. The process is analogous to that used in scientific refereeing of a paper. This will show whether the food items are appropriately described, that the samples are representative, and identify the analytical methods and the quality criteria used by the authors of the data. It will also identify whether the modes of expression used are compatible with the mode chosen and which interchange 'tags' are appropriate. This stage also provides the documentation for the food item and the procedures used in generating the data.

This stage may generate queries that would need to be asked of the analysts and, in the case of published material, may require further information from the authors.

Construction and Evaluation of an Archive of All Data

This involves bringing together all the existing data for the composition of a single food item. The values are then scrutinized for consistency and where the data sets are of sufficient size, a range of statistical analyses is conducted on the data. Before this analysis, however, it is necessary to bring all the data into the agreed modes of expression. Visual inspection and statistical analysis serve to identify discordant values and characterize them as statistical outliers [11]. This will require an iteration to the data source to ensure that the discordant values have been correctly entered, that the food item is correct, that the analytical methods are compatible, and that the modes of expression are concordant. If these criteria are satisfactory, then it will be necessary to consider whether the variation observed is real, that is, it occurs naturally, or whether the extreme outlying values should be discounted. There is no objective way to make this decision because the analytical values for some trace elements, such as chromium, originally showed very high variance. As they have been improved by collaborative studies and the use of standard reference materials, the accepted values now correspond to the outlying low values obtained initially. The correct course of action at this stage is to implement studies of the analytical methodology and to arrange for the collection and analysis of a number of additional samples. The data base compiler may have to use the mean or median value with qualification as an interim measure until this further work is available for scrutiny.

At this stage some numerical checks for internal consistency can be carried out, for example, the recovery of the food item by summing the proximate constituents and comparison of sums of amino acids with protein values and fatty acids with total fat values.

Constructing and Evaluating a Comprehensive Set of Data

This stage may correspond to the final stage of the data base preparation and deals with the evaluation of the assembled values for all the food items. This is most conveniently carried out by food groups because the primary objective is to assess the mean (or median) values, preferably with statistical measures of variability, for food items of a similar type. This may show inconsistencies in variance or the values themselves, which require iteration to the archival collection to confirm the values or with the data source to see whether the identity of the food or some feature of the sampling and analysis is the cause of the inconsistency. Where foods have been processed or cooked in different ways, comparisons of the data should show no inconsistencies, for example, in the composition of the dry matter between raw and prepared items.

Construction and Evaluation of the User Data Base

In many cases, the comprehensive data base will meet the needs of users but in others different specifications may require the preparation of separate user data bases. This would be the case where the users require single values for nutrients or a restricted range of nutrients, for example, the users may require vitamin A value as retinol equivalents, which will require combining carotene and retinol values. Other users may require restricted data sets either in respect of food items and/or nutrients. If the earlier evaluations are satisfactory then these user data bases can be prepared automatically from the comprehensive data base. Some users may be content to have data that do not meet the rigorous criteria suggested earlier, and lower quality data could be brought forward from the archival collection. This would be justified for many very uncommon foods where an analysis has been carried out on one occasion and where the further analysis of the food is impracticable, such as a food collected in the wild or food that is rare and expensive, for example, caviar or champagne.

Compilation of the Data Base

The processes suggested for evaluating the data culminate in the effective preparation of the data base itself because assembling the data provides the best context for data evaluation. The final stages involve developing the documentation of the data sources, deciding on the actual descriptions of the food items that are to be included, and documenting the analytical methods. Some authors suggest that quality codes [12] should be assigned to each value but the use of numerical codes for attributes that are not real numbers has its dangers in that the codes may be used as if they were real numbers, whereas in many cases they are subjective assessments that are difficult to validate. The true test of quality of a value is the accuracy with which it predicts the composition of another equally representative sample of the food.

We do not propose to discuss the structure of the data base in the computer sense because this is outside the remit of the chapter. However, the general principle that the data base structure and its associated management system should be designed to meet the users' needs remains the first factor to consider. The development of programs to manipulate the data similarly must follow users' requirements, and because these programming needs will change and develop with time, a data base system that permits manipulation of the data cells as part of its management system as in-built software is highly desirable.

Validation of the Data Base

It is useful to consider how data bases should be evaluated because this forms the final test of the system's quality. Various tests of data bases have been suggested based on comparisons of the output, usually by comparing calculations of lists of foods such as would make up a diet [13]. These comparisons are useful in testing consistency of operation but frequently have an element of circularity because many data bases use the same data source. Comparison of the predictive accuracy for calculating the composition of analyzed dietary mixtures is a more critical test because the composition of the analyzed diet has defined levels of accuracy. Even so, this remains a demonstration rather than a proof of accuracy, and the natural variability of foods will set limits to the level of accuracy that can be achieved. Nevertheless, many such comparisons would reduce the probability that the agreement has occurred by chance. In practice, few data bases in existence have been subjected to any kind of validation of their predictive accuracy; yet, their values are widely used and often given spurious levels of accuracy when the results of calculations using them are cited [14]. The limitations on validation mean that the major source of reliability in a data base must be built in during its construction.

Documentation of the Data Base

The final stage in the preparation of the data base involves developing documentation that describes it and its operation to potential users and provides a reference for the users that assists in using the data base. This should include not only the correct usage but also an account of its limitations and improper usage.

In the case of a printed data base made available as a series of tables, the documentation takes the form of a textual introduction to the tables and gives an account of the structure of the data base and the criteria used in compiling the data. Other information on analytical methods, sources of data, and additional tables where the information is too limited to be included in the main tables may be given as appendixes [15]. The introduction contains details of the limits of accuracy of the data base and a list of caveats for the user. Some kind of index is also essential to enable the user to find specific food items or nutrients. The index is necessary because the printed format places physical limits on the size of the tables so that the nutrients are distributed among different tables and pages in the book.

In the computerized data base the size of the food item by nutrient matrix of the data base itself is virtually unlimited and search facilities are usually very rapid. The users require an instruction manual of some kind and ideally this should form part of the computerized system and be accessible during use via 'help' screens or some similar system. The documentation concerning analytical methods, descriptions of samples, and sources of data can be directly linked to the main data base matrix. The aim should be to ensure that all information the user requires to use the data base correctly is accessible via suitable software. This will call for the development of nutrient and food information files that document their respective topics. Many users will need software that identifies specific food items from which the compositional data can be used for foods that are missing and for missing nutrient values.

The preparation of a data base depends on achieving a good working relationship between the users and those responsible for constructing the data base. Ideally, as the data base passes through its various stages, the users should be consulted and the choices made should have their agreement. If this approach is followed the final data base should be very close to being a useful nutritional tool. However, like all tools, only regular use will prove whether the compilers have achieved their objectives, and the process of review and revision will lead to the evolution and refinement of the data base into a really effective instrument.

References

1 Greenfield H, Southgate DAT: Guidelines for the production, management and use of food composition data: An Infoods manual (in press 1991).

2 Greenfield H, Southgate DAT: A pragmatic approach to the production of good quality food composition data. ASEAN Food J 1985;1:47–58.

3 Southgate DAT: Guidelines for the Preparation of Tables of Food Composition. Basel, Karger, 1974.

4 Kelly A (ed): Nutritional Surveillance in Europe: A Critical Appraisal. Wageningen, Euro-Nut, 1986.

5 Cameron ME, Staveren WA (eds): Manual on Methodology for Food Consumption Studies. Oxford, Oxford Medical Publications, 1988.

6 Cheek PR (ed): Toxicants of Plant Origin. Boca Raton, CRC Press, 1989, vol 1–4.

7 Arab L, Wittler M, Schettler G: European Food Composition Tables in Translation. Berlin, Springer, 1987, pp 132–154.

8 Haendler H: Methods of identifying data units for retrieval purposes as applied in an international data bank system for feed analyses; in Proc Int CODATA Conf, Jerusalem 1985. Amsterdam, Elsevier Science, 1985, pp 401–404.

9 Klensin JC, Feskanich D, Lin V, et al: Identification of Food Components for INFOODS Data Interchange. Tokyo, United Nations University, 1989.

10 Ministry of Agriculture Fisheries and Food: The British Diet: Finding the Facts. Food Surveillance Paper No 23. London, HMSO, 1988.

11 Youden WJ, Steiner EA: Statistical Manual of the Association of Official Analytical Chemists. Arlington, Association of Official Analytical Chemists, 1975.
12 Exler J: Iron content of foods. Home Economics Research Report No 45. Washington, US Department of Agriculture, 1982.
13 Hoover LW: Computerized nutrient data bases. 1. Comparison of nutrient analysis systems. J Am Diet Assoc 1983;82:501–505.
14 Paul AA, Southgate DAT: Conversion into nutrients; in Cameron ME, Van Staveren WA (eds): Manual on Methodology for Food Consumption Studies. Oxford, Oxford Medical Publications, 1988, pp 121–144.
15 Paul AA, Southgate DAT: McCance and Widdowson's – The Composition of Foods. London, HMSO, 1978.

David A.T. Southgate, PhD, Head, Nutrition Diet and Health Department,
AFRC Institute of Food Research, Norwich Laboratory, Colney Lane,
Norwich, NR4 7UA (UK)

Simopoulos AP, Butrum RR (eds): International Food Data Bases and Information Exchange. World Rev Nutr Diet. Basel, Karger, 1992, vol 68, pp 49–93

Validated Data Banks on Food Composition: Concepts for Modeling Information

Max Feinberg, Jayne Ireland-Ripert, Jean-Claude Favier

Centre Informatique sur la Qualité des Aliments, Paris, France
Institut National de la Recherche Agronomique (INRA)
Institut Français de Recherche Scientifique pour le Développement en Coopération (ORSTOM)

Contents

Introduction

Analytical Chemistry and Food Science

Over the past few years, there has been renewed interest in food nutrient composition data in many countries. This comes after several decades of relatively few developments in analytical food chemistry but of an expansion of analytical chemistry to become an independent discipline. Between 1930 and 1950, food chemistry was an important part of analytical chemistry and highly developed in agricultural research laboratories and universities. Thereafter, interest grew for the feedstuff industry and feeds analysis. At the same time, analytical chemistry found a very large field of applications in the pharmaceutical and the fine chemicals industries. This evolution is reflected in the changes brought to the acronym of the AOAC: in the beginning, it was translated as the Association of Official Agricultural Chemists; now it is known as the Association of Official Analytical Chemists.

When dealing with the storage and treatment of analytical data, it can also be seen that more importance has been given to animal rather than to human nutrition. The considerable work undertaken by the INFIC (International Network of Feed Information) [1] clearly illustrates that the interest for an international interchange of data on feedstuffs came earlier than for foodstuffs. The INFIC had already proposed an international coding system for feeds in the 1960s [2, 3], whereas the comparable network for foods, called INFOODS (International Network of Food Data Systems) [4, 5], was only created in the 1980s.

Nonetheless, the growing interest in food nutrient composition data has led to the creation of food composition data banks, containing updated analytical data, in several countries. This tendency has even recently given rise to a new publication, the *Journal of Food Composition and Analysis,* which presents results of research prompted by the implementation of these data banks [6].

These food composition data banks are 'official', national – maybe soon international – and are different from the numerous computer programs that have been developed using previously published food composition tables. It must be said that the production of such software preceded this new movement toward the creation of food composition data banks. For instance, more than 200 food composition computer programs already exist in the United States but, in general, they all use data compiled from the US Department of Agriculture (USDA) Handbook No. 8 [7]. An annual National Nutrient Data Bank Conference presents up-to-date information on such products based on USDA nutrient data [8]. In comparison, more than 40 such computer programs are available in France [9].

This development can partly be explained by the obsolescence of data edited in earlier composition tables. New foods are produced every year, and improved analytical techniques give more detailed or more precise determination results. But this argument alone is insufficient, as the elaboration of a national data bank is an expensive, nonprofitable operation. Simple financial balance is impossible to reach most of the time, and institutes involved in this work have to be largely supported by government funding.

Deep motivations with regard to food composition are thus of another nature, and resorting to a data bank is a means, not an end. There has been flagrant inertia in adopting dependable physicochemical methods for quality control of foods, despite the fact that the exchange of food products has intensified over the past years. In 1985, in France, Germany, Italy, Japan, the United Kingdom, and the United States, food export and import involved more than 10% of the total local international trade, and this proportion is constantly increasing. Consequently, objective and simple means of controlling food quality must be defined. The food composition data bank is one of the tools in this strategy. Moreover, considering the rapid increase in the consumption of processed foods and the growing role of catering in our society, governments may quite soon need to propose nutrition policies. Here again, food composition data banks have an important role to play.

At the same time, interlaboratory studies organized on food analysis often prove that precision in analytical methods is frequently not achieved. Leading laboratories in different countries have been shown to produce widely different values for macronutrients in common foods [10]. The means to improve this situation – better standardization of methods and reference materials of certified nutrient concentration – are only beginning to be organized at the European level by the Community Bureau of Reference [11, 12]. Many expert meetings also take place in the framework of the Codex Alimentarius to propose adapted methods. The studies involved in this question are extensive, and food composition data banks may help to provide useful information in this respect.

Data Banks as Tools for Food Science

This paper is not a review in the usual sense of the word, if by review we mean an article that brings together summaries of research findings and assessments of research methods. Rather, it is an introduction to what is meant by food composition and an illustration of some of the ways computing methodology can be successfully applied to structuring complex chemical information. Much material has been taken from meetings of task groups and not from published material. Many examples are taken from the results obtained in our laboratory in developing our own food composition data bank.

Computer science is not only the science of computer architecture. It also indicates fruitful and scientific ways to organize and improve information management. It is well known that nutrition is deeply related to social and cultural habits; this, in turn, influences food science. Computer science provides possible methodological approaches to propose a more rigorous way of reasoning.

In the context described above, many governmental organizations, like the French Ministry of Agriculture, have decided to create national data banks on food composition. In France, this task was entrusted to the Centre Informatique sur la Qualité des Aliments (CIQUAL), in the Netherlands to the CIVO Institute of TNO, and in the United Kingdom to the MAFF Food Service Division. Their principal aim is to collect data obtained from various laboratories performing nutrient determinations on foods consumed in the country where the data bank is situated. From these data, reference values are elaborated that can serve either scientific purposes, such as dietetics or nutrition policy, or commercial and administrative purposes to control food quality. Analytical results can be said to constitute the raw material of data banks on food composition. We have chosen to describe the scientific background behind these projects to explain the concepts used to implement such data banks and permit consistent and validated data interchange between data compilers.

Such an approach, which consists of using a data bank as a tool of information validation, is not unique in the history of data banks. Basically, all data banks can be conceived according to two very different approaches:

(1) The first approach consists of storing collected information in a random manner; this information is then enhanced in value by the power of the data base management system, which helps to create a compatibility in the data.

(2) The second assumes a preliminary structuring of the information; the compatibility of data is verified previously. This amounts to choosing a standardized model for data, which can then be validated when entered into the bank.

Considering the goals at stake in food chemistry, the second approach is scientifically preferable and must therefore be chosen. Moreover, it justifies the efforts required to develop a national food composition data bank. However, efforts are still necessary, as to this day none of the existing food nutrient data banks is accessible to on-line users.

Data Modeling

Information Flow in Food Chemistry

A preliminary goal is to define the information inputs: which data will be stored and what volumes will be managed. According to the general functioning plan chosen, the data are obtained from public and private analytical laboratories specializing in the analysis of human foods. These laboratories are the data generators. A possible alternative would be to directly associate a laboratory to the data bank instead of collecting data from many laboratories. Although this solution seems simpler, it has drawbacks. The investment for such a laboratory would be very high and underused, and the existence of such a laboratory is no guarantee for the collection of accurate and precise data; mastering many sophisticated analytical techniques is not always an easy task and the efficiency of such a laboratory might not be satisfactory.

Next, it is necessary to define the operations to which these collected data will be submitted. They must be put into standard form, entered, stored, and then processed by the data bank software; the people involved in these operations are the data compilers. These procedures can be defined as an aggregation of raw data. Aggregated data are then submitted to referee committees in charge of determining their validity from an analytical as well as from a nutritional point of view.

To perform the task of validation, it is therefore important always to be able to trace the origin of data, before and after their aggregation. To render operational this principle of functioning, a certain number of preliminary conditions must be fulfilled:

(1) The creation of a descriptive coding system for foods, based on the internal needs of food composition data banks and harmonized at the national and international levels to facilitate data exchange.

(2) The elaboration of a laboratory network to furnish data, carrying out interlaboratory tests and collaborative studies, which guarantee the quality of the analytical results validated by the referee committees.

(3) The definition of statistical algorithms and methods for the aggregation of data to compute reference values that are compatible from a statistical point of view. This task must not be neglected as optimal sampling conditions are scarcely ever reached, and it is impossible to define a correct sampling design when no knowledge is available on data dispersion.

At the other end of this path are the users. Users of numerical data on foods have variable needs because they belong to various professions with multiple interests. Nutritionists are involved in epidemiological studies or clinical investigations. Dietitians need data to compute suitable diets. Food manufacturers have to produce new food products, consistent with consumer needs or tastes. Government officials require standards to elaborate new regulations or define the conditions of the food quality control and labeling.

The overall demands are extremely variable, from a global appreciation of common foods, representative of a certain type of food consumption or consumer behavior, to the research of detailed data on specially selected foods or raw material for clinical or experimental studies [13–15]. For instance, an epidemiologist working on the relationship between cancer and nutrition may wish to evaluate the influence of 'eating grilled bovine meat twice a week over 20 years' and needs a composition estimate for the beef. This is very different from the requirements of a food manufacturer who wants to elaborate a new low-calorie deep-frozen meal containing certain cuts of beef. The former needs a global composition of some kind of representative bovine meat, which may be computed as an average value of different meat cuts weighted by the statistics of their consumption; this is more a mathematical model than an actual food. On the other hand, the food manufacturer will need to compare the compositions of several meat cuts to optimize the recipe according to the energy content or the cost. They both have a common need for data that are representative of foods available on the national market, but they do not give the same meaning to the word 'bovine meat'.

Rather than data users, these professionals can be defined as data interpreters. Consumers are also highly interested in accessing food composition data. Generally, this information must be trivialized at this level because nutritional education is not yet common in many countries.

Thus, information flows from its source – the generators – through the organizations involved in structuring data – the compilers – to its redistribution among the different types of users – the interpreters (fig. 1). To

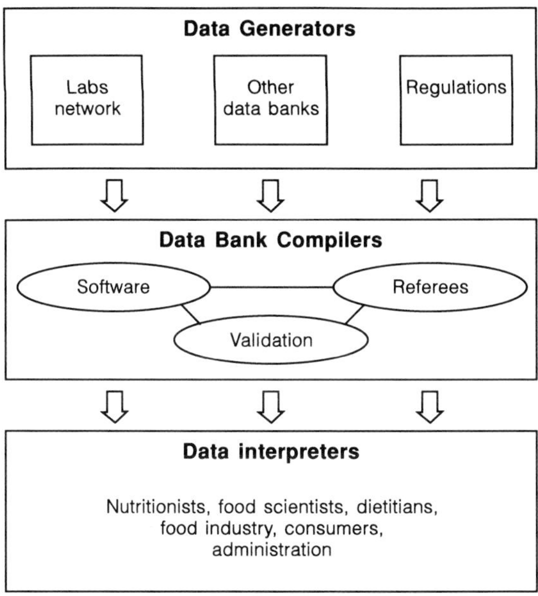

Fig. 1. Global plan of information transit.

satisfy the diversity of users' requirements, the means of data access or consulting must be made flexible, that is, capable of being adapted to a variety of requests. These means of access can be, for example, specialist books [16–19] but can also be magnetic tapes to facilitate the incorporation of data in other systems or automatic means of on-line access and data transfer through computer networks.

Entity-Relationship Model

As stated earlier, the data stored in a modern food data bank have to be modeled to be certified in accordance to this model. To perform this task, a rigorous line of reasoning must be followed. It may appear at first as an inconvenience, but the benefit will be to give strict definitions to the concepts handled in the data bank. For instance, one of the first steps is to build a precise catalog of all types of data to be collected and to standardize their definitions.

The computer certification program of the food composition data bank created by the CIQUAL is called the Répertoire Général des Aliments or REGAL. It has the responsibility to furnish users not only with factual information of the chemical composition and the nutritional value of foods but also with documentary information on food regulations [20, 21]. The data bank was developed according to a certain data design, incorporating a certain number of methodological concepts. The REGAL data bank can be used as an example to explain and illustrate these concepts.

One classical method to carry out the logical analysis of information is based on the principle that the real world can be perceived as a collection of objects, called entities, and relationships between these objects. An 'entity-relationship' model is thus created to structure data for computer applications [22]. This type of model provides fairly flexible structuring capabilities and allows one to specify data constraints explicitly. It has gained acceptance as an appropriate data model for data base design and is widely used in practice.

An entity is an object that exists and is distinguishable from other objects. For example, in the context of food composition data management, bread and rice are defined as two objects of the same type. An entity set is a set of entities of the same type. A data base thus contains a collection of entity sets, each of which includes any number of entities of the same type. In the case of data relating to nutritional composition, one entity set is obviously food and another is constituent.

Distinction among entities is accomplished by associating each entity with a set of attributes that describe it. For instance, possible attributes for food are the food name and the food group; for constituent these can be the constituent name and the unit used to express the result. For each attribute there is a set of permitted values, called the domain of that attribute. The domain of the attribute food name might be the set of all text strings of a certain length. Table 1 gives several examples of constituent entities. A list of entities presented in such a format is also called a table.

A relationship can exist among several entities. By convention, a relationship between two entities, entity 1 and entity 2, will be noted < entity 1 & entity 2 > in the text and as a diamond shape in diagrams, where entity sets are represented by square boxes and attributes by ellipses. A recursive relationship only involves one entity. The best example is trying to describe a recipe with the entity-relationship model. A recipe is composed of a list of several food items. A simple qualitative description consists of relating one food, called the recipe, to other foods in its composition, called the ingredients. Thus, a reflexive relationship is built, written < food & food > following our convention. To be more quantitative, it is interest-

Table 1. Examples of *constituent* entities

Constituent name	Unit
Alanine	mg/gN
Alanine	g/kg
Alginates	g/kg
Alpha-tocopherol	% tocopherol
Alpha-tocotrienol	% tocopherol
Aluminum	µg/kg
Amylopectin	g/kg
Amylose	g/kg
Amylose (monosaccharide equivalents)	g/kg
Arachidic acid (20:0)	% fat
Arachidonic acid (20:4 n-6)	% fat
Ash	g/kg
Barium	µg/kg
Biogenic amines	mg/kg
Biotin	µg/kg
Butyric acid (4:0)	% fat
Butyric acid (4:0)	g/kg
Caffeine	mg/kg
Capric acid (10:0)	% fat
Capric acid (10:0)	g/kg
Cholesterol	% sterol
Cholesterol	g/kg
Dry matter	g/kg
Edible part	kg/kg
Energy	kcal/kg
Energy	kJ/kg
Fiber (acid detergent method)	g/kg
Fiber (neutral detergent method)	g/kg
Fiber, total dietary	g/kg
Glycerides, total	g/kg
Glycogen	g/kg
Glycogen (monosaccharide equivalents)	g/kg
Magnesium	mg/kg
Manganese	µg/kg
Organic selenium	µg/kg
Polyunsaturated n-3/polyunsaturated n-6	p. 100
Polyunsaturated n-3 total/polyunsaturated total	p. 100
Polyunsaturated n-6 total/polyunsaturated total	p. 100
Protein, total	g/kg
Zinc	mg/kg

ing to store the amount of each ingredient used. Thus, a relationship may also have descriptive attributes.

In general, a relationship is an association between several entities. In the model of a food composition data bank, it is clear that there is a very important relationship between food and constituent. Each time an analysis is performed on a given food for a given constituent, these two objects must be related to store numerical data through the contains relationship noted <food & constituent>. For each contains relationship, we expect to find, as attributes, the average value, the minimum value, the maximum value, and the number of samples analyzed (fig. 2).

The exact number of entities to which another entity can be associated via a relationship is called its cardinality, and this depends on the model used. Its value is important during the logical structuring of the data. Between two entities A and B, different types of relationships can exist, according to the cardinality:

- (1,1) where an entity in A is associated with at most one entity in B, and an entity in B is associated with at most one entity in A.
- (1,n) where an entity in A is associated with any number of entities in B, but an entity in B can be associated with at most one entity in A.
- (n,1) where an entity in A is associated with at most one entity in B, but where an entity in B can be associated with any number of entities in A.
- (m,n) where an entity in A is associated with any number of entities in B and vice versa.

In the above example, contains is a binary relationship with a cardinality (1,n). Each entity in the set food is associated with any number of entities in the set constituent.

Many relationships in a data base are not binary but can involve more than two entity sets. For example, information on food composition is obtained from different laboratories. We must then create a new entity set, laboratory, which stores the name of the laboratory where the constituent was analyzed in the studied food product. Contains would then be a ternary relationship: <food & constituent & laboratory>. In this case, it is necessary to define cardinalities for each entity set couple: for food toward constituent it is (1,n), for laboratory toward constituent it is also (1,n), and for food and laboratory it is (m,n) as, in general, laboratories are able to analyze several kinds of food.

When data modeling, distinguishing between entities and relationships is not always easy. In a data base, individual entities and relation-

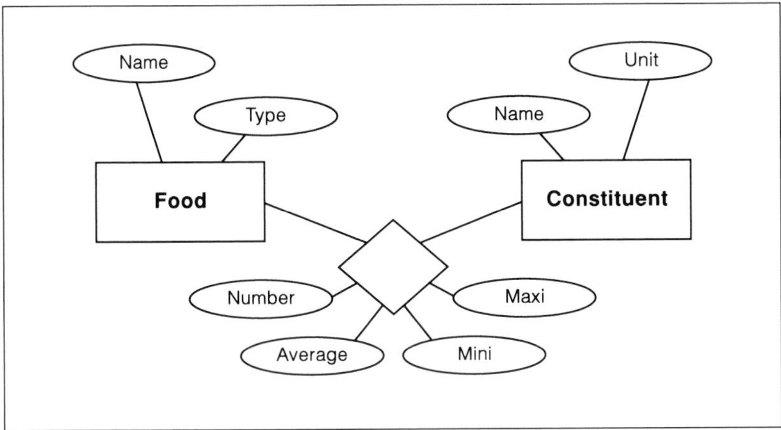

Fig. 2. 'Contains' modeled as a binary relationship.

ships are distinguished by their attributes. To make such distinctions, a superkey is assigned to each entity set. The superkey is a set of one or more attributes, which, taken collectively, allow us to identify an entity in the entire set. By means of the superkey, each entity will be unique in the data bank, although each occurrence of a relationship may have duplicates.

For example, a nonrepeated code number, added to the attributes of each food (or each constituent), is sufficient to distinguish one food product (or one constituent) from another. Thus, the superkey for the food (or constituent) entity set is such a code. Using the example of the REGAL data base, we prefer to use a numerical code instead of using the name of the food product as a superkey, as the same entity may have several names, and several entities may have similar names. It is also customary to use a numerical superkey because it is more easily manageable by the available data bank management systems and it is shorter. But there is no established rule about this, and the only criterion for a superkey is its uniqueness.

Because superkeys can be defined by various attributes, the same information may be modeled in different ways. Thus, to model chemical information, it may be preferable to define an entity set called analytical measurement having a superkey that is a chronological registration code and treat the analytical information as a ternary relationship of the type <analytical-measurement & food & constituent> (fig. 3).

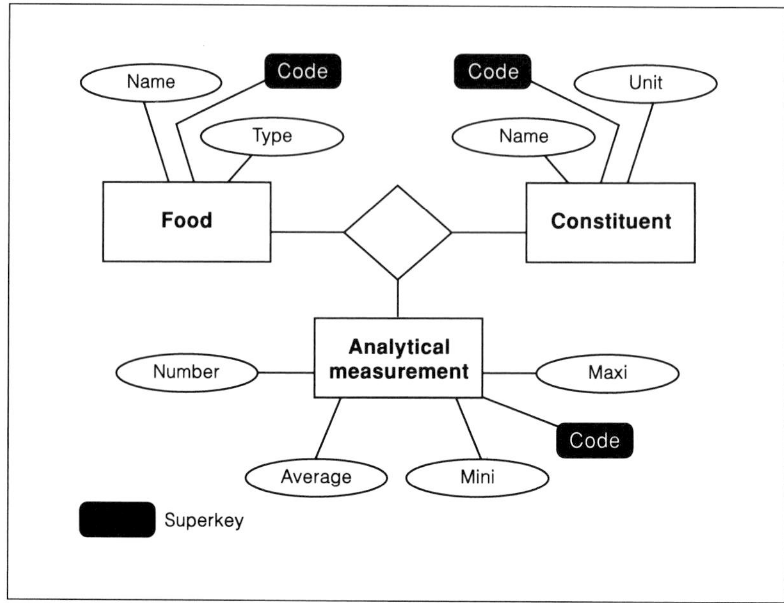

Fig. 3. Analytical measurement modeled as a ternary relationship.

Once the model has been constructed, the next step is to define which type of file organization handles this information best. The choice is generally a compromise between flexibility and ease of access.

Information on Food Composition

Food nutrient information can be put into two kinds of data. One kind of data consists of analytical determinations. These are in constant evolution, changing when new methods appear, destined to be permanently improved and more representative of actual foods. These are called factual data and are meant to be processed often for computational purposes. On the other hand, these data are only interesting if related to foods and constituents.

The second type of data is of a very different nature. It is documentary information, related to semantics, food science, or analytical chemistry. This information is used to build the setting of the factual data. It is also the task of the data bank compilers to collect and standardize environmental data. They are obliged to propose common vocabulary and common reference procedures. To do this, they need to aggregate various kinds of data.

Food regulations and biological science are very important in this context, even more so when data are exchanged between different countries. This is evident if we consider food classifications. Each language has typical words used to name food categories that can be difficult to translate. For example, the English word 'berries', used to group fruits as different from a botanical point of view, such as strawberries and blackberries, cannot be translated with one word in French. On the other hand, the French word 'charcuteries' is often improperly translated as 'cooked pork meat' [23] because it includes raw and preserved pork meat in addition to processed meat products like sausages and pâtés. Thus, it is necessary to find specific ways to design a system for retrieval of this type of data.

Environmental Data

Food Products

Within the context of REGAL, we considered as a food any substance that can be digested and/or assimilated by human beings. Although apparently trivial, this definition is taken in a larger sense than that accepted by many laws, French regulations among others, because it includes additives in addition to raw materials, transformed products, prepared dishes, and recipes. In this case, air is not to be regarded as a food but water is.

The only attribute of the entity set food is its superkey. This consists of a 5-digit code, which allows 99,999 different foods to be identified; the choice of this number will be justified below (see Coding Foods). Besides this superkey, each entity is distinguished by a mandatory name. One important problem is to obtain, whenever possible, a translation of each food name. As we intend to collect information about tropical foods and recipes, it is impossible to give preference to one language. Moreover, it is often requested to enter the name of the same food (or constituent) in several languages. Therefore, an entity language was created with its own internal code and name. The food name is associated with the entity set language throughout the relationship <language & food> (fig. 4). The entity set language enables the storage of as many synonyms as necessary. This enables the treatment of data on foreign or regional foods (from laboratory results or from literature) and the publication of multilingual composition tables.

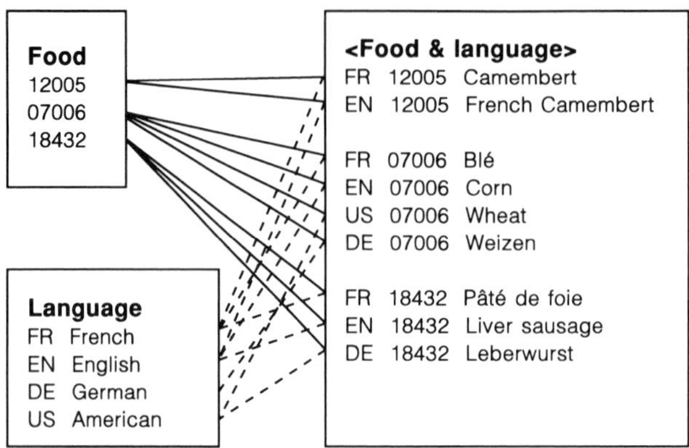

Fig. 4. Entities and relationship 'How to name a food product in different languages'.

Although the chosen coding system is elaborate, it may be insufficient to describe complex or typical foods. To resolve this problem, the concept of recipe was introduced. It is a very convenient way of describing a food product made from other foods, such as a prepared dish. This also explains why the food definition was extended to additives, as they are frequent ingredients of complex foods. This composition of a recipe entity is registered in a relationship <food & food>, which contains the quantity of each food ingredient contained in the food recipe. A recipe can be used to describe a food but also to calculate nutrient composition values from the components. In practice, each item of this relationship has three attributes: the food code of the food recipe; the food code of the food ingredient, and its amount in grams or percentage. A quick scanning of the first column of this table (food recipe codes) reveals which foods are registered as recipes.

Unfortunately, a simple name does not always allow precise management of composition data or an exchange of information with others. Many synonyms, homonyms, and homographs exist even between closely related languages. For example, the UK 'endive' corresponds to the US 'chicory' and vice versa. Names of what were once specific foods, such as 'cheddar cheese' or 'gruyère', have become trivialized and are now used for a variety of products. Foods that are ethnic or national in origin often differ in various countries because of the necessity to comply with local regulations and consumer tastes. For example, Danish marinated herrings are not the same when purchased in Denmark and in other countries because food manufacturers adapt their compositions to local tastes. Although 'bread' can be easily translated by 'pain' or 'brot' when using a dictionary, this does not mean that French or German bread is made the same way as English or American bread. Clearly, the name of a food does not reflect its chemical composition.

In addition to its name, it is thus necessary to have a clear description of each food product, as scientific as possible. With this aim in view, a system of descriptive food codification is being devised between foreign data banks to exchange 'pertinent' information [24]. Notably, a Committee on Data for Science and Technology (CODATA) Task Group on a Systematic Nomenclature for Foods in Numeric Data Banks has been formed to study the descriptive codification of foods. The CODATA [25] is an interdisciplinary Scientific Committee of the International Council of Scientific Unions (ICSU) that seeks to improve the quality, reliability, management, and accessibility of data of importance to all fields of science and technology.

Coding Foods

Requirements for a Coding System

The design of a coding system must be the result of several considerations: simplicity, accuracy, and convenience. In the case of food composition data, we must consider the following three constraints:

(1) The number of food items (or entities) to be coded.

(2) The possibility of relating this code to the scientific description of foods.

(3) The ability to use the code for retrieving nutritional information.

It is easy to estimate the number of different foods, as defined above, that are to be stored in a data bank. This number is between 5,000 and 10,000. In the REGAL data bank, we decided that the entity set food would contain about 5,000 entities, and this is why a unique 5-digit code is assigned to each food product. In no way must this food code be considered as an international coding system; it is only an internal, sequential identification number and may be compared with the registration number used in the documentation CAS data bank (Chemical Abstracts) for each chemical compound [26].

More than 150 food composition tables are listed in the INFOODS directory of food composition tables [27]. In general, the foods are described simply, by common names, supplemented by taxonomic Latin names where relevant; occasionally descriptions include information on the method of cooking or conservation. Foods are listed alphabetically or by conventional food group (e.g., dairy, meat, grain, vegetable), but no two countries appear to agree on these classes or what should go into them, especially when mixtures are concerned.

The EUROCODE coding system was developed as an attempt to universalize such food groups. However, EUROCODE is only a loosely structured listing of food names that provides no means of capturing and encoding other information. To implement such a simple hierarchical code, it is necessary to define an a priori rigid structure for the items to be coded. For foods, this consists of classifying products in families, subfamilies, sub-sub-families, etc. For instance, to code 'stewed beef', the first step is to define the food type; the first field would thus be M for meat (beef, veal, pork, mutton). The second place (subgroup of meat) is 1 for beef, thus the code becomes M1. The third and fourth places proposed for M1 pieces of meat would be, respectively, 8 (dishes) and 8 (stewing meat). The entire code for 'stewed beef' is then M188. On the other hand, 'beef stew' would be coded Y184: Y (animal dishes), 1 (meat dishes of beef), 8 (stewing meat), 4 (stew). However, some of the ambiguities and contradictions have perhaps been resolved in the latest edition of EUROCODE [28, 29].

This hierarchical coding method does have the advantage of being simple and relatively easy to use. The code obtained can be used as a superkey. Positions of each figure or letter can also be used to decipher a food as well as to retrieve the code number of a food item. Unfortunately, such a coding system presents three major drawbacks, which render it badly suited to our purposes:

(1) To be precise enough for foods, the number of hierarchical levels must be great. If a coding system provides between 7 and 10 levels (is this a reasonable limit?), the simplicity of the method is no longer quite evident. Moreover, if only numerical, such a code allows for 10^7–10^{10} food items to be coded. Considering the 5,000 expected food entries, its efficiency is thus very poor.

(2) The simple tree-shape structure of this coding system makes it very difficult to modify when new foods appear. Any changes imply a complex set of transformations, dangerous for the integrity of the data bank. In addition, the number of subfamilies is variable, and the development of the system may lead to blocking situations.

(3) The most critical problem is that foods cannot be easily classified, It has already been stressed that, depending on the country, grouping may be very different. Moreover, some foods may belong to several classes. In mathematical terms, food groups can be defined as fuzzy sets. It may be difficult to decide to which food group some complex foods belong. For instance, should stewed beef with carrots be classed as a 'composite dish with meat' or as a 'composite dish with vegetables'? Even simple foods can be difficult to classify; for example, is a croissant a 'cake' or a 'bread'?

These problems were already encountered when developing the INFIC coding system for animal feeds [30]. There it was decided to use a different method, better adapted to the reality of food science and based on a descriptive coding system and a thesaurus structured in facets.

The LanguaL Coding System

The food coding system that was chosen for REGAL is called LanguaL (Langua Alimentaria). It is based on the Factored Food Vocabulary (FFV) system developed more than 8 years ago by the US Food and Drug Administration (FDA), NIH university scientists and USDA for classifying food products for information retrieval purposes [31–34]. It is presently used in the following food data banks:

(1) FDA Total Diet Study (quarterly analysis of typical market basket, on residue values of pesticides, toxic elements, nutrient elements, and chemicals).

(2) FDA Scientific Information Retrieval and Exchange Network (food additives and regulatory information on 3,200 food products).

(3) Food Component Research Database of the National Cancer Institute (NCI), based on the Nutrient Database for Standard Reference (5,000 foods, nutrient analytical data) of the USDA [35].

(4) Centre Informatique sur la Qualité des Aliments, France. In France, the LanguaL system has been translated into French for the French food data bank and for the Algerian and West African food data banks, in collaboration with the corresponding countries.

LanguaL has also been used to code foods by the Health and Welfare Food Composition Data Bank of Canada, the National Food Agency of Denmark, and the Ministry of Agriculture, Fisheries and Food of the United Kingdom.

However, the LanguaL system cannot be considered as totally satisfactory. A major problem is that it was first developed in the United States,

Table 2. Facets used for *food* coding

Facet (code and name)	Cardinality: definition
A Product type	1,1: Family or group of foods defined by common consumption, functional or manufacturing characteristics
B Food source	1,1: Animal, plant or chemical source from which the product or the primary ingredient is derived
C Part of plant/animal	1,1: Anatomical part of the plant or animal from which the food product or its major ingredient is derived (meat, milk, root, sugar)
Z Adjunct characteristics	0,3: Quality criteria (label, meat cut, plant maturity) and other characteristics (type of crust, beverage mix)
E Physical state or shape	1,1: Physical state of food product as a whole (solid, liquid)
F Extent of heat treatment	1,1: Extent the food has been modified in processing by the application of heat (raw, cooked)
G Cooking method	1,2: Process by which a food product is cooked (broiled or grilled, deep-fried, cooked with steam)
H Treatment applied	1,12: All physical or chemical treatments applied to the product or its major ingredients; also describes additives and ingredients
J Preservation	1,2: Primary method used to prevent microbial and enzymatic spoilage
K Packing medium	1,2: Substance in which the food is packed for preservation and handling and/or palatability
P Consumer/dietary group	1,3: Group for which the food product is marketed (regular diet, low fat)

so many terms used in the thesaurus are very closely related to North American nutritional habits and legislation. Other deficiencies derive from the nontaxonomic classification of plants and the impossibility of coding parts within parts of organisms (e.g., cod liver oil). Modifications and improvements may soon appear in the framework of the international CODATA Task Group. On the other hand, the principle of this coding system makes modifications easy even in existing data banks. Globally,

Table 3. Facets used for *sample* coding

Facet (code and name)	Cardinality: definition
M Container or wrapping	1,3: Defines the main container material, the container form and the liner, lids and end materials
N Food contact surface	1,3: Material or materials which actually touch the food product
Q Establishment	0,1: Origin of the sample: type of establishment (factory, store ...)
R Geographical origin	1,1: Where the food was produced (country, region)
S Storage conditions	1,2: Length and conditions of storage preceding the analysis
T Production period	0,1: Year or period of production

LanguaL remains a good starting point for development of a truly international and flexible faceted thesaurus.

As constructed, LanguaL is a thesaural system using faceted classification; each coded object is described by a set of standardized terms, regrouped in facets. Each facet represents a set of characteristics that affects the nutritional quality and/or safety of a food product, such as product type, food source, preservation, or cooking methods (tables 2, 3).

LanguaL contains more than 2,500 standardized terms. A short excerpt from the thesaurus (table 4) shows that terms are structured as son terms and father terms. The descriptors within each factor are arrayed in a hierarchy from broader to narrower terms to facilitate retrieval and aggregation of data. Up to 10 hierarchical levels can be used. For instance, one might search specifically for 'soft-ripened cheese' (or 'soft-ripened cheese' plus 'cured or aged') or more broadly for 'cured cheese' or most broadly for 'cheese or cheese products', or one might aggregate consumption of all foods having 'cow' or 'curd' as source. The hierarchical arrangement also displays the vocabulary in a logical way to facilitate indexing and retrieval.

From a wider point of view, the coordination of the recipe entity set of REGAL and the LanguaL coding system allows us to build a true network among foods, raw materials, additives, and recipes. Such a model can easily be the starting point of an information model adapted to expert

Table 4. Excerpts from the LanguaL thesaurus

A. Product type	Dairy product
	Cheese or cheese product
	Natural cheese
	Cured cheese
	Hard grating cheese
	Hard cheese
	Semi-soft cheese
	Soft-ripened cheese
	Mold-ripened cheese
	Uncured cheese
B. Food source	Animal used as food source
	Meat animal
	Cattle
	Cow
	Goat
C. Part of plant or animal	Part of animal
	Milk or milk component
	Cream or cream component
	Curd
	Milk
	Whey
E. Physical state, shape or form	Solid
	Divided or disintegrated
	Whole
	Whole, natural shape
	Whole, shape achieved by forming
	Whole, shape achieved by forming, thickness <0.3 cm
	Whole, shape achieved by forming, thickness 0.3–1.5 cm
	Whole, shape achieved by forming, thickness 1.5–7 cm
	Whole, shape achieved by forming, thickness >7 cm
H. Treatment applied	Food modified
	Microbially/enzymatically modified
	Enzymatically modified
	Clotting agent added
	Lactose converted
	Fermented/modified, complex process
	Cured or aged
	Lactic acid – other organism fermented

systems. Compared with the monohierarchical coding system (e.g., EUROCODE), the faceted thesaural system appears to be highly flexible and well adapted to the development of new food products. There are no limitations to handling new products or new categories of products when they appear. Its major drawback stems from it needing considerable international collaboration to be efficiently implemented. This goal has already successfully been attained for the faceted INFIC coding system, which is widely used throughout the world and is a reference point for all concerned in the composition of animal feeds.

Control of Coding Compatibility

In such a system, a 'reasonable' limit to description precision must be defined. For example, describing 'bread' and describing 'French baguette bread bought in a Parisian bakery in 1989' do not represent the same amount of coding effort. The more precise the description, the more extensive it is; extensive coding needs more time for input and requires more space in the data bank computer. To take this limit into account, it is useful to introduce an entity set called sample. The LanguaL coding system can also be used to describe the sample.

The sample represents, in some ways, the portion, or the aliquot, of food actually analyzed in a laboratory. Moreover, this entity is very useful for managing the different analytical measures collected by the data bank compiler. Each occurrence of sample is simply identified by a chronological identification number used as a superkey and has a complementary attribute consisting of a commentary written in free text for additional information not described by the coding system. Food is related to sample by means of a (1,n) relationship (fig. 5). It must be noted, however, that this concept is not an obligation of the LanguaL system but was introduced for the needs of the REGAL data bank. The LanguaL method is described in detail in the next paper by Dr. Thomas C. Hendricks.

To control and standardize the faceted coding that differentiates the entities food and sample, we introduced a thesaurus cardinality table that contains the coding system rules in an appropriate form. To understand its importance in maintaining the coherence of the coding system, it is interesting to see what this table contains. It consists of five columns:

(1) The facet code, composed of a letter from A to Z.

(2) The name of the facet as reported in tables 2 and 3.

(3) An integer value representing the minimum number of times each facet may be used in a codification.

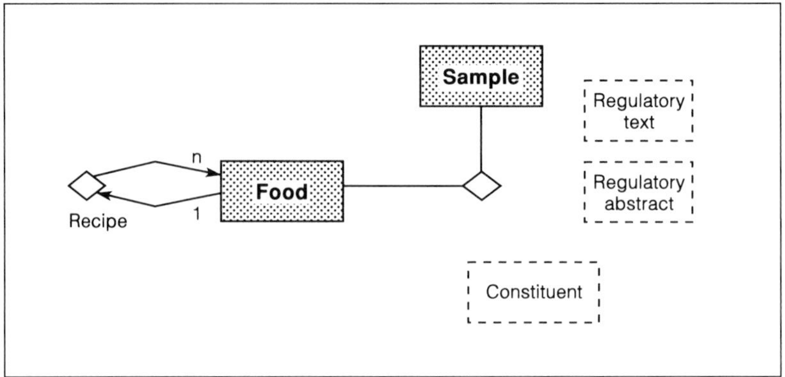

Fig. 5. The <food & sample> and <food & food> relationships.

(4) An integer value representing the maximum number of times each facet may be used in a codification (see the column called cardinality in tables 2 and 3).

(5) A logical flag indicating whether a facet is used to describe a food product or a sample.

When the minimum value is set to 0, this means that the use of the facet is optional. According to the cardinality table used for REGAL, for example, from 1 to 12 terms of the H facet (treatment applied) can be used to describe a food whereas the Q facet (establishment) is optional for sample description, as the lower cardinality is 0. Using the facet identification letter and the cardinality table, it is always possible to know if a food or sample description is complete or if, on the contrary, the limits have been reached.

With such a method of control it would be possible, for example, to increase the cardinality of facet A and use several A-letter terms, so that beef stewed with carrots could be described as a 'composite dish with meat' as well as a 'composite dish with vegetables'. However, in this case it is perhaps simpler to create a new term such as 'composite dish with meat and vegetables'.

A food is described by means of standardized terms or descriptors contained in the 11 facets listed in table 2; a sample is described by means of standardized terms (descriptors) of the facets listed in table 3. According to the cardinality table, between 10 and 29 descriptors can thus be used for coding a food product. With this system, a food analyzed as bought and a prepared food do not have the same descriptors, especially for the facets E, F, and G, and thus do not have the same food identification code.

Thesaural Structure

All the standardized descriptors, whether they are used for food products or for samples, are entities of the same entity set called glossary. Each record of this entity set possesses a superkey composed of the letter identifying its facet, followed by a 4-digit number. This 4-digit code number is internal to the data bank and is 'transparent' to users who utilize LanguaL descriptors to retrieve information on food composition. Moreover, in the REGAL data bank, the glossary table includes the French and English names of each descriptor. When the system is extended to other languages, it will be possible to create a relationship <glossary & language> containing the name of each descriptor as an attribute.

Associations between a food product, or a sample, and each LanguaL descriptor constituting its definition are made through the relationships <food & glossary> or <sample & glossary>. One danger of any documentation system such as LanguaL is the 'drift' that may occur in the meaning of thesaural descriptors. To avoid this and also improve retrieval, each descriptor can be associated with a standardized definition, based on legal texts or examples (see below).

To widen or narrow requests, a hierarchy was created among descriptors of each facet. Table 4 shows some broader descriptors that are the fathers of narrower descriptors. This hierarchy is managed by a recursive relationship <glossary (father descriptor) & glossary (son descriptor)>, depending entirely on the glossary entity set. Broader and narrower descriptors can thus be indicated for each descriptor and modifications in the hierarchy easily handled. A simple algorithm allows the complete tree diagram to be rebuilt from this relationship [36].

Some examples of food and samples coding are given in tables 5a–c. In particular, descriptions of French and American bread are compared. The differences observed in the description of bread can illustrate part of the discrepancies among tables. Several studies were performed to compare the composition of the 'same' foods in composition tables from different countries [37]. The variability observed, if partly due to systematic bias, such as analytical methods and sampling, is also simply caused by the fact that comparison was not made on the same foods but on foods colloquially given the same name. Even data of good quality can be a source of error if they are derived from foods that are not clearly defined.

Constituents

In the context of REGAL, we considered as a constituent any quantitative chemical, physical, nutritional, or sensory characteristic. This definition thus includes oleic acid, density, percent of refuse, color, etc. At present, purely qualitative data, like flavor description or technological appreciation, are not treated in the data base. On the other hand, concentrations of additives or metabolites of additives can also be included in the list of constituents. In this case, the special position of additives should be noted, which can be considered, according to the circumstances, as food products or as constituents. For example, it is clear that alpha-tocopherol measured in carrots by high-pressure liquid chromatography is a constituent; on the other hand, the additive, classified E.307 in European nomenclature, or synthetic alpha-tocopherol, is a food because it can be an ingredient in a recipe and because it may itself contain several analyzed constituents.

Table 5a. LanguaL coding of the 'Camembert de Normandie' from raw milk

Description of the food product		
A	0138	Soft-ripened cheese
B	1201	Cow *(Bos taurus)*
C	0245	Milk curd
E	0105	Whole, shape achieved by forming, thickness 1.5–7 cm
F	0003	Not heat treated
G	0003	Cooking method not applicable
H	0107	Lactic acid – other organism fermented
H	0298	Clotting agent added
H	0289	Cured or aged 2–4 weeks
J	0003	No preservation method used
K	0003	No packing medium used
P	0024	Human food, no age specification, regular diet
Z	9001	Mold rind
Z	0087	Controlled name
Description of a sample		
M	0177	Wood box
M	0173	Paper wrapper
N	0039	Paperboard or paper
Q	0004	Manufacturing establishment
R	0250	France
S	0003	Not stored
T	0488	Production April 1988

Table 5b. LanguaL coding of the French bread (from CIQUAL)

Description of the food product		
A	0178	Bread
B	1418	Hard wheat *(Triticum aestivum)*
C	0208	Seed or kernel, skin removed, germ removed (endosperm)
E	0105	Whole, shape achieved by forming, thickness 1.5–7 cm
F	0003	Complete heat transformation
G	0005	Baked or roasted
H	0256	Carbohydrate fermented
J	0003	No preservation method used
K	0003	No packing medium used
P	0024	Human food, no age specification, regular diet
Description of a sample		
M	0003	No container or wrapping used
N	0003	No food contact surface present
Q	0002	Retail establishment
R	0250	France
S	0001	Storage conditions unknown
T	0689	Production June 1989

Table 5c. LanguaL coding of the American white bread (from FDA)

Description of the food product		
A	0178	Bread
B	1418	Hard wheat *(Triticum aestivum)*
C	0208	Seed or kernel, skin removed, germ removed (endosperm)
E	0151	Solid
F	0003	Complete heat transformation
G	0003	Cooking method not applicable
H	0256	Carbohydrate fermented
H	0136	Sugar or sugar syrup added
H	0194	Enriched
H	0181	Iron added
H	0216	Vitamin B added
J	0001	Preservation method not known
K	0003	No packing medium used
P	0024	Human food, no age specification, regular diet
Description of a sample		
M	0001	Container or wrapping not known
N	0001	Food contact surface not known
Q	0001	Establishment unknown
R	0840	USA
S	0001	Storage conditions unknown
T	0001	Production period unknown

The entity set constituent also needs a superkey. Recent studies show that about 400 constituents with a nutritional role have been registered in different nutrient data banks or composition tables [38]. Thus, a simple 3-digit number can be used as an internal constituent code. For REGAL, a 5-digit code was chosen to allow extended room for storage of analytical data on contaminants and metabolites. Thus, the number of entries for the constituent table may go above 10,000.

Like food products, constituent may be named in different languages through a relationship <constituent & language>. Because data can be furnished or transcribed in different units, it is also useful to introduce an entity set unit. Its attributes are an internal unit code and a name (table 6). Moreover, a <unit & unit> relationship contains a factor of conversion to facilitate conversions among units. The basis of this factor is the kilogram, as the international unit system MKS (meter-kilo-second) was chosen for REGAL. However, table 6 can be extended to different unit systems. For instance, the <unit & unit> relationship is not convenient when transforming the amino acid concentrations expressed in percent of total nitrogen to mg/kg concentrations; in such cases, it is necessary to use a formula (see below).

To clarify editions such as composition tables, one can use an entity set constituent group, which classifies constituents in useful homogeneous lists, such as vitamins, minerals, proximate analysis, etc. The name of a constituent group in a chosen language is accomplished through a relationship <language & constituent-group>. The constituents belonging to each group are identified by a relationship <constituent & constituent-group>.

Regulatory Texts

Another category of environmental data is of different nature. It involves the regulations related to foods. Food regulations vary from country to country but always aim to define a set of criteria for foods to help protect the consumer's health, prevent fraud, and ensure honest trade. Regulatory information is therefore a prerequisite to any marketing of a food product. However, this information is very complex, abundant, and sometimes confusing [39, 40]. Is it necessary to incorporate such information in a food nutrient data bank?

In developing the data model for food composition, it is clear that at least three types of data processed by the bank are affected by regulations:

Table 6. The entity set *unit*

Code	Denomination	Code	Denomination
1	kg/kg	7	kJ/kg
2	%	8	mg/gN
3	g/kg	9	% fatty acids
4	mg/kg	10	% sterol
5	µg/kg	11	% tocopherol
6	kcal/kg	12	–

(1) The standardized descriptors stored in the glossary entity set require standardized definitions to be used pertinently. This standardization was partly carried out by people responsible for regulations. For instance, the LanguaL descriptor 'flavoring or flavor enhancer' (A0300) is defined as follows:

> 'Substance added to a food to supplement, enhance, or modify the original taste and/or aroma of a food or any of its ingredients without imparting a pronounced characteristic taste or aroma of its own' [41].

(2) Many foods have imposed legal constituent or nutrient concentrations. This is particularly important for contaminants or additives but also for nutritional compounds. For instance, table 7 shows the differences in EEC and US regulations for milk fat content of different categories of milk.

(3) Several administrative bodies have already developed coding systems at the national or international level, for example, the European coding system for imported goods [42]. It is interesting to build a 'bridge' between these nomenclatures and food composition. This can be done through a relationship between external administrative codes and the internal food code used in the data bank.

Unfortunately, relationships of these legislative texts with food composition data are complex, as they can involve one food product and one constituent (relation 1,1), a group of foods and one constituent (relation n,1), or one food product and a group of constituents (relation 1,n). References to legislative texts pertaining to food law can be included in an entity set called regulatory text. Its attributes are an internal code used as a

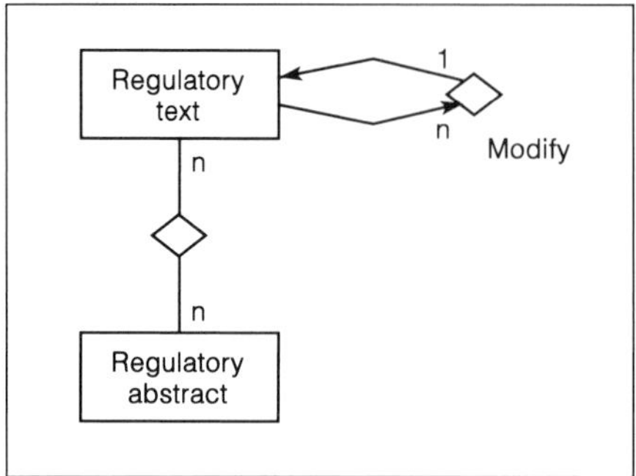

Fig. 6. The relationship between regulatory-text and regulatory-abstract.

Table 7. Example of different food regulations

Food	Europe[1]	USA[2]
Whole milk	≥3.5% fat	≥3.25% fat
Half-skimmed (EEC) or low-fat (US)	1.5–1.8% fat	0.5, 1, 1.5 and 2% fat
Skimmed	≤0.3% fat	≤0.5% fat

[1] EEC Regulation 1411/71 of June 29, 1971.
[2] Code of Federal Regulations 21, chapter 1.

superkey, the nature of the text when necessary (law, decree, regulation code, etc.), the title of text, and its reference.

However, because of their complex relations to food composition, these texts are not directly useful. To manage such complexity, an efficient way would be to create another entity set, called regulatory-abstract, which contains a summary of one of several texts pertaining to one food and one constituent (fig. 6).

The relationship <regulatory-abstract & food & constituent> (fig. 7) allows access to summaries established in function of the type of food product and/or type of constituent and information on legally admissible concentrations in the composition of a food product. This two-level model takes into account a great variety of situations. In addition, a regulatory abstract can be written from one or several legislative texts and can concern one or several food products and one or several constituents. Because a regulatory abstract is related to the legislative texts on which it is based –

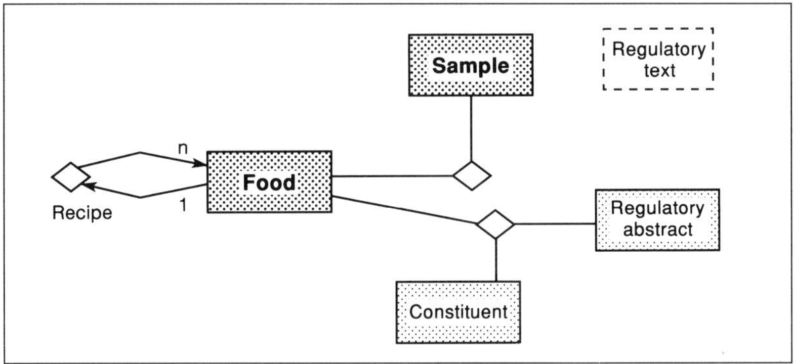

Fig. 7. The relationship <regulatory-abstract & food & constituent>.

relationship>regulatory-text & regulatory-abstract> – these references are always available for additional information.

Moreover, if a new standard modifies an already existing one, a relationship <regulatory-text & regulatory-text> points to the texts concerned (fig. 6). This relationship is very similar to the relationship between father and son descriptors in the glossary entity set. Once again, information is organized as a network that is not 'frozen' but easily modified and updated. Furthermore, a link is created between each descriptor of the glossary entity set and its standardized definition based on regulatory texts; it is the relationship <glossary & regulatory-text> whose sole attribute is this definition.

Finally, other food codification systems can be stored in a unique entity set external code with the following attributes: a code identifying an external organization (e.g., USDA, Customs Registration Number, etc.) and the food code given by that organization. A relationship <food & external-code> yields these codes via the internal food code.

Nevertheless, the difficulties encountered in structuring regulatory texts must not be neglected. Regulatory information is, in general, very language-dependent. Fortunately, retrieval of references is simple, thanks to the many documentary data banks existing in the field of regulation and jurisprudence. The application of computer science to regulatory information is a recently developed field, where application of computer science to expert systems can be successfully applied (e.g., expert systems to calculate retirement pensions for employees and veterans).

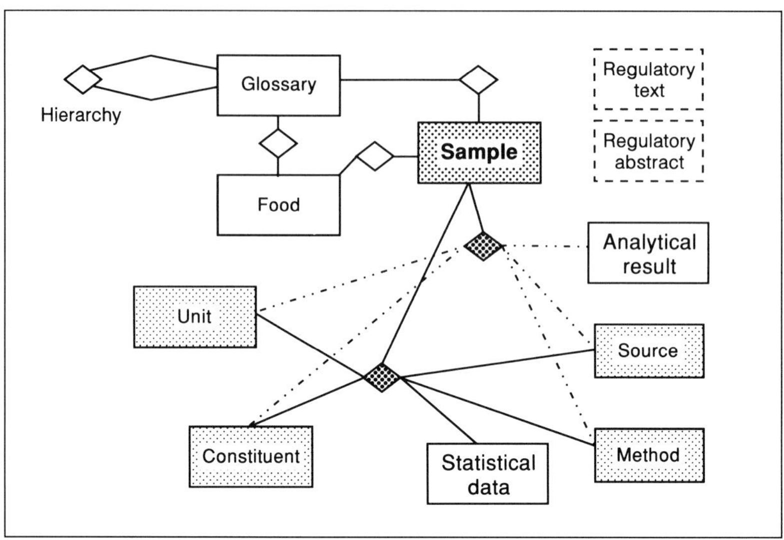

Fig. 8. The relationships between statistical-data or analytical-results and the other entity sets of the model.

Sources of Data and Analytical Methods

To facilitate the validation of composition data, these must be identified with codes to trace their origin through the entity set source and the analytical method used to obtain them through the entity set method. Source contains attributes giving either the name and address of a laboratory that performed the determination or the bibliographical references where the data were published. Sometimes the 'laboratory' can be a group of laboratories and the publication can be a handbook; this extension has no influence on the nature of the information, which remains compatible. In REGAL, the source superkey is a code composed of either the letter <B> for a bibliographical reference or the letter <L> for a laboratory plus a 4-digit number. Complementary attributes can be used to identify accredited laboratories or reliable publications. The role of this information is to help the referee committee make decisions. How this entity is related to chemical data is described below (fig. 8).

In the same way, another entity set method indicates the analytical method used by the source. In REGAL, the attributes of this entity set are a 5-digit internal code and a succinct description of the method used. It is not intended to manage an exhaustive manual of analytical methods but to allow a simple and global reference to analytical techniques. For example, if it is important to know what analytical method was used to determine dietary fiber, experts should know if a result was obtained by the AOAC method or by another. On the other hand, this concept could possibly be extended to the management of a real analytical manual. This would require, for instance, the addition of an entity set method-manual containing the protocol used for the specific analysis. This text would be related to the method and constituent code by means of an adequate relationship.

Documentary Access and User Interface

LanguaL codes, consisting of one letter and four digits, can be used directly for retrieval purposes; the coding system is designed for that purpose. Simple programming, using boolean equations, allows any kind of combinations of codes. The logical operators AND, NOT, and OR give a very flexible way of selecting dynamic groups of foods. However, there are presently more than 2,500 descriptors, their number is regularly increasing, and descriptor codes may be esoteric for most users. As an on-line aid, lists of keywords can be created and stored in specific entity sets. The definition of adapted relationship renders documentary access to environmental information possible:

(1) An entity set glossary-keyword can be developed to make retrieval of LanguaL descriptors simpler. According to the different languages used, several such entity sets may be necessary, such as French-glossary-keyword or English-glossary-keyword.

(2) For analytical methods, the entity set method-keyword, and for chemical data sources, the entity set source-keyword, assume the same role.

Each of these XXX-keyword entity sets is, of course, associated with its corresponding entity by a specific relationship. This allows the user to have recourse to a documentary type of research.

Factual Data

Data Collected from Laboratories

In the case of food nutrient data banks, the factual data can be restricted to chemical data. As far as environmental data are precisely defined, it is easy to define a chemical datum as the determination of one constituent, in one sample, obtained by one source using one method and expressed in a single unit. However, depending on the context, it can be useful to differentiate different kinds of chemical data. Two types of data from laboratories are integrated in the REGAL data bank:

(1) Raw analytical results, corresponding to one measurement carried out on one sample, for one constituent. A complete set of data is collected.

(2) Composite results, from a series of measurements carried out on one food product and for one constituent. Generally, several statistical parameters have already been calculated by the laboratory, such as average or standard deviation, and are included in the data sent to the data bank compiler.

The distinction between isolated analytical results and previously calculated statistical data led to the creation of two separate entity sets: analytical-result and statistical-data. Each of these entity sets has several similar attributes: a two-element superkey composed of the sample number and a chronological data registration number, plus a counter registering the number of times a datum has been used afterwards in any computation performed by the data bank compiler. This information is completed by the analysis date and an average value. What differentiates the analytical-result entity from the statistical-data entity is that a standard deviation, high and low values, and the number of measures can also be specified for the latter.

The role of the counter is important during computation of reference data. Several algorithms are available that incorporate or discard values depending on statistical criteria (see below). Data that have never been used can thus be eliminated from the data base because they can be considered as outliers, not representative of the couple (food, constituent). This counter is a flag indicating the relevancy of data. Figure 8 illustrates the relationships between statistical-data or analytical-result and the other entity sets described above.

Although only quantitative data are required, frequently important analytical measurements are expressed qualitatively, for example, 'trace', 'observed absence', 'below detection limit', etc. [43]. For instance, it is well known that vitamin B_{12} is absent from vegetable foods; it is important for food technologists or biochemists to indicate that copper is present as a trace element in some oils causing potential peroxidation. Storing this kind of information using 'special' values, such as a negative value or to mix text with numeric data, may lead to confusion when the data are used in computations or may create difficulties in programming.

Table 8. The entities of the *status table,* as used in REGAL

Code	Rank	Denomination
1	1	measured or calculated
2	2	trace
3	3	probably present (not measured)
8	8	not present
9	9	unknown

A solution is to introduce a status table composed of an internal code, a name, and a factor of classification among states (table 8). Each numerical datum in analytical-result and statistical-data is accompanied by a complementary flag containing a code value of status table indicating whether it is quantitative or qualitative. Before performing any mathematical operation between pieces of data, each status code is verified, and if the code is different from '1' (true numerical data), the corresponding data are not used, as not to warp the results, for example, in a calculation of the average. On the other hand, the factor of classification of states permits statistical methods based on rank, such as calculation of the median (see below).

Reference Values

It must be recalled that the main goal of a modern food data bank consists of creating reference values used for scientific or commercial purposes. These data are 'imputed' or 'aggregated' from raw or composite data sent by the laboratory network. They must be permanently recalculated when new data are introduced. This represents a lot of work and justifies the use of automated computation methods easily accessible on computers. However, to avoid abusive aggregation of data, the data bank compiler must be cautious in performing this task. This can be achieved by suitable modeling of information.

Computation of Reference Values

Aggregated-data are data created by the data compiler by calculation. They can be obtained either from a series of analytical-results and/or statistical-data or from other aggregated-data. They represent the real added value of the data bank. Each aggregated datum is identified by a superkey formed by its food, constituent, and unit codes. This entity set has the following attributes: a central value (average or median), standard deviation, high and low values, the number of measures used to obtain the result, the date of data aggregation, and the aggregation method chosen.

Figure 9 shows the relationship between aggregated-data and the other entity sets. Figure 10 indicates the flow diagram for computing aggregated-data. The box marked 'formula completed' represents the search for information necessary for the aggregation of data. This search itself depends on the type of calculation and formula. Given their importance, the different aggregation methods are described below:

(1) Calculation of the average or the median of a series of analytical results and/or statistical data. This calculation can be carried out automatically, or most often, interactively with or without weighing of results. It necessitates a preliminary selection of data to be taken into account. As it is fundamental to be able to trace the basic data used for aggregation, counters stock this information in the analytical-result and statistical-data entities (number of times a result or datum has been used to obtain a reference value).

(2) Formulas and extrapolation from other aggregated-data. For example, a value for energy is obtained from the concentrations of proteins, fats, and carbohydrates of the same food product. This kind of calculation is very frequent, and many mathematical models have been published. Sometimes it is even used as the basis of a regulation, such as the 6.38 coefficient used to obtain the protein content of dairy products from total nitrogen.

(3) Recipe, from other aggregated-data, when the food product is composed of several food ingredients.

Given the risk of using formulas too automatically, this question has to be examined attentively. The existing formulas in human nutrition belong to four general types of mathematical models:

(1) Product or ratio, involving one single food. This is the standard way to convert units (e.g., joules to calories) or to impute a nutrient from one or two other constituents. All formulas of this type form the formula-1 entity set.

(2) Addition or shift, used to add or subtract two complementary constituents (e.g., moisture and dry weight) or to shift the value for one constituent by a constant value. All formulas of this type form the formula-2 entity set.

(3) Summation is applied to an entire set of constituents, such as amino acids or fatty acids. This allows one to aggregate separate nutrients

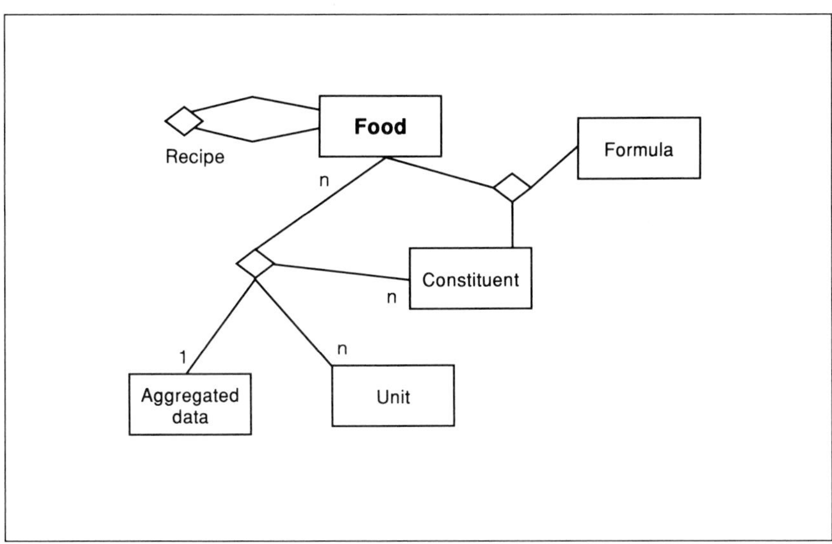

Fig. 9. Relationship between aggregated-data and other entity sets.

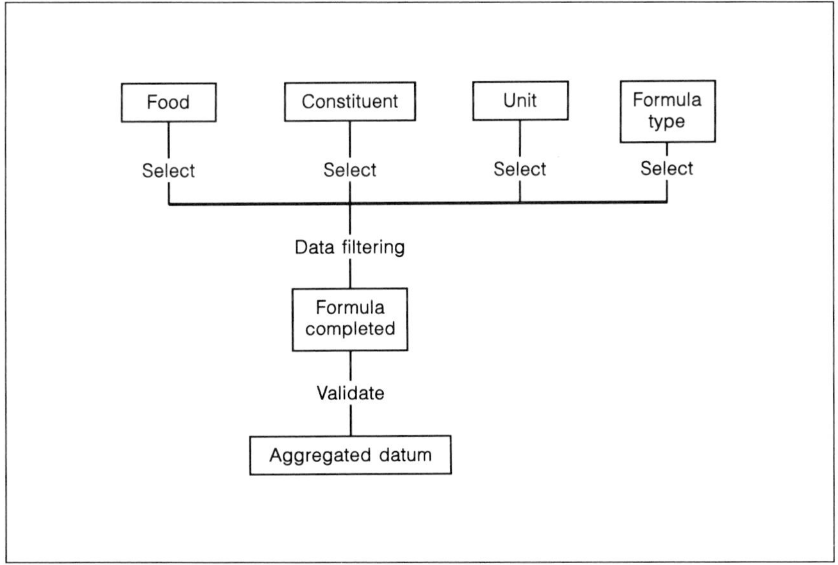

Fig. 10. Flow diagram for computing aggregated data.

of similar nutritional effect and build a new, more convenient criterion. All formulas of this type form the formula-3 entity set.

(4) Extrapolation or transfer can be compared with the product operation but involves more than one food. All formulas of this type form the formula-4 entity set.

The superkey of aggregated-data corresponds to a triplet made from the identification codes of one food, one constituent, and one unit. By convention, this can be represented by the notation (A/C/U), where A symbolizes the food code, C the constituent code, and U the unit code. Then each type of formula can be noted as follows:

Table 9. Examples of formulas

Product (formula 1)	
Proteins from nitrogen concentration	$[A \mid \text{proteins} \mid g/kg] = k_p \cdot [A \mid N_{total} \mid g/kg]$ [1]
Energy unit conversion	$[A \mid \text{energy} \mid kJ] = 4.184 \cdot [A \mid \text{energy} \mid kcal]$
Fatty acid concentration, g/kg	$[A \mid FA \mid g/kg] = k_{FA} \cdot [A \mid FA \mid \%FA_{total}] \cdot [A \mid \text{lipids} \mid g/kg]$ [1]
Amino acid concentration from nitrogen	$[A \mid AA \mid g/kg] = [A \mid AA \mid mg/gN] \cdot [A \mid N \mid g/kg]/1{,}000$
Vitamin A in different units	$[A \mid \text{vitamin A} \mid \mu g/kg] = 0.3 \cdot [A \mid \text{vitamin A} \mid IU/kg]$
Vitamin A from β-carotene	$[A \mid \text{vitamin A} \mid \mu g/kg] = k_b \cdot [A \mid \text{carotene} \mid IU/kg]$ [1]
Addition (formula 2)	
Dry matter and humidity	$[A \mid \text{dry mat.} \mid g/kg] = a_0 + a_1 \cdot [A \mid \text{humidity} \mid g/kg]$ [2]
Humidity and dry matter	$[A \mid \text{humidity} \mid g/kg] = a_0 + a_1 \cdot [A \mid \text{dry mat.} \mid g/kg]$ [2]
Sum (formula 3)	
Defatted dry matter (DFDM)	$[A \mid DFDM \mid U] = [A \mid \text{dry mat.} \mid U] - [A \mid \text{lipids} \mid U]$
Sum of saturated fatty acids	$[A \mid FA_{sat} \mid U] = \text{sum}\,([A \mid FA_{sat(i)} \mid U])$ [3]
Sum of monounsaturated fatty acids	$[A \mid FA_{mono} \mid U] = \text{sum}([A \mid FA_{mono(j)} \mid U])$ [3]
Vitamin A activity	$[A \mid \text{vitamin A} \mid \mu g/kg] = [A \mid \text{retinol} \mid \mu g/kg] + [A \mid \text{carotene} \mid \mu g/kg]/a_1$ [4]
Extrapolation (formula 4)	
Concentration in the edible part of food	$[A_{\text{edible part}} \mid C \mid U] = [A_{\text{as bought}} \mid C \mid U]/a_0$ [5]
Yield of a treatment	$[A_{treated} \mid C \mid U] = a_0 \cdot [A_{origin} \mid C \mid U]$ [5]

[1] The values of k_p, k_{FA} and k_b depend on the food.
[2] $a_0 = 1{,}000$ and $a_1 = -1$.
[3] FA_{sat} and FA_{mono} are the sums of saturated and monounsaturated fatty acids. $FA_{sat(i)}$ and $FA_{mono(j)}$ are the sums of individual fatty acids with $1 \leq i \leq 30$ and $1 \leq j \leq 12$.
[4] $a_1 = 6$ (2 for dairy products).
[5] a_0 is % of the edible part or the effect of the treatment, such as loss of vitamins during cooking.

Formula-1: product or ratio

$$(A \mid C \mid U) = a_0 \cdot (A \mid C_1 \mid U_1) \cdot (A \mid C_2 \mid U_2)^{a1}$$

with $-1 \leq a_1 \leq +1$

Formula-2: addition or shift

$$(A \mid C \mid U) = a_0 + a_1 \cdot (A \mid C_1 \mid U_1)$$

Formula-3: sum

$$(A \mid C \mid U) = \text{sum}(a_1 \cdot (A \mid C_i \mid U_i), \text{ with } 0 \leq i \leq n$$

Formula-4: extrapolation or transfer

$$(A \mid C \mid U) = a_0 \cdot (A_1 \mid C \mid U_1) \cdot (A \mid C_1 \mid U_2)^{a1} \cdot (A_1 \mid C_2 \mid U_3)^{a2}$$

with $-1 \leq a_1, a_2 \leq +1$

Each entity of the formula-1 sets contains, as attributes, the coefficients or the codes necessary for its application. Thus, for a formula-1 entity, there is a formula number, the multiplicative constant a_0, the power a_1, and the constituent codes C_1 and C_2, as well as the corresponding unit codes U_1 and U_2; for a formula-2 entity, there are only a_0, a_1, C_1, and U to be furnished.

To control or select formula type, it is necessary to employ another entity set, called formula, which is used to catalog these different types of formulas. For each entity in formula, there is a brief description of the recorded formula, a code indicating the formula type (1, 2, 3, or 4), and a numerical formula code. Table 9 gives several examples of calculations using each type of formula. When a reference value is obtained by formula, this information is stored through a relationship <formula & food & constituent & unit>. In this way, only one kind of relationship is necessary to access any kind of formula.

Calculation by recipe (table 10) can be formalized with the same conventions as formulas, and we obtain:

Recipe:

$$(A \mid C \mid U) = \text{sum}(a_i \cdot (A_i \mid C \mid U_i))$$

Validation of Reference Values

A set of management rules must be integrated to the software to avoid abusive calculations (insufficient amounts of data, for example) or errors. The first criteria take into account the status code of data, but others are based on statistical theory to define the most exact method of calculation or best adapted method of expressing a result according to the selected data. This can be done checking the probability density function of the data set. Unfortunately, the amount of data is generally too small to use powerful statistical tests.

Table 10. Examples of computations with recipes

Simple recipe	$[A \mid C \mid U] = \{[A_1 \mid C \mid U] \cdot \text{weight}(1) + \ldots + [A_n \mid C \mid U] \cdot \text{weight}(n)\} / \{\text{weight}(1) + \ldots + \text{weight}(n)\}$
Composition of an average food	$[A \mid C \mid U] = \{[A_1 \mid C \mid U] \cdot \text{part\%}(1) + \ldots + [A_n \mid C \mid U] \cdot \text{part\%}(n)\} / 100$

A possible alternative consists of using nonparametric statistics, such as the median and percentiles instead of the arithmetic mean and its confidence interval, to give the central value and an estimate of the data dispersion [44, 45]. This fundamental computational step is entirely performed by the software. It is regrettable that no real standard exists in this field. One further task of data interchange committees should be to give guidelines on this question to improve the quality of data validity.

Each entity in the aggregated-data set contains, aside from the data calculated as above, three additional attributes:

(1) The number of individual data used for the calculation.

(2) The method of calculation (average, median, formula, recipe).

(3) A flag indicating whether the data have been certified by the scientific committee.

Using the network established between the different pieces of information, the scientific committee can work from a set of objective data:

(1) Analytical methods used to determine the analytical results or statistical data and, if necessary, the origin of the data generator (laboratory or literature).

(2) The type of algorithm used for calculation and the number of measurements involved. Even if the number of measurements is unknown (for instance when data are taken from literature), this information is stored as it is important for validating the final aggregated data.

(3) The sample descriptions. For instance, the data and the region of production of the food product can be very important when dealing with vegetable oils. The free text commentary, accompanying the sample description, can help explain abnormal distribution of measurements or outliers.

Moreover, repeatability and reproducibility computed from collaborative studies can be stored and used to describe analytical determinations and reference values. An entity set precision – containing the repeatability, the reproducibility, the number of laboratories involved, and the date of execution – can be related to the other parts of the data model through a relationship <precision & method & source & constituent>. The source links shows that such a datum can be obtained from literature or from a collaborative laboratory study. In this connection, the Reference Material Committee (REMCO) of the International Standardization Organization (ISO) has developed a data bank (COMAR) to assist chemists in finding the reference material needed for collaborative studies; four national coding centers collect certified reference materials from national producers and merge the information into this international data base [46–49].

However, repeatability and reproducibility computed from collaborative studies constitute a global method of evaluating an analytical method, and there are too few results for each couple (food, constituent) studied. It is necessary to make some adjustments when dealing with nutrient data bases.

Software Implementation

Data Bank Management Systems

One major benefit of using the entity-relationship model to describe information is that it is directly connected to commercial software called Relational Database Management Systems (RDBMS). A RDBMS lets the user visualize all information – entity sets and relationships – as a collection of tables made of vertical columns (attributes) and horizontal rows (individuals). (See Srivastava and Butrum, p. 20.)

In this respect, a well-known standard is called Structured Query Language (SQL) [50, 51]. SQL is sometimes said to be a user-friendly language as it is nonprocedural compared with languages such as COBOL. By non-procedural, we mean that the user only needs to know how – and not where in the computer system – the data are stored.

Let us consider the example illustrated in figure 4, 'How to name a food product in different languages'. The tables involved are defined as follows using SQL data dictionary language [52]:

```
% Table containing the internal food code
TABLE FOOD;
    food-code: INTEGER, NOT NULL; % internal food code
    KEY (food-code);
END TABLE FOOD;
```

```
% Table containing languages and their codes
TABLE LANGUAGE;
    lang-code: CHAR(2), NOT NULL; % ISO code of language
    lang-name: CHAR(15), NOT NULL; % name of language
    KEY (lang-code);
END TABLE LANGUAGE;

% Table containing the relationship <food & language>
TABLE FOOD-LANGUAGE;
    food-code: INTEGER, NOT NULL; % internal food code
    lang-code: CHAR(2), NOT NULL; % ISO code of language
    food-name: CHAR(70), NOT NULL; % name of food
    KEY (food-code, lang-code)
END TABLE FOOD-LANGUAGE;
```

To create a structure to hold data, the first step is to define a table, that is, name its columns and state their data types. For example, the table LANGUAGE contains two columns: lang-code (the ISO code for languages, composed of 2 characters) and lang-name (the name of the language, which may use up to 15 characters). 'NOT NULL' signifies that the column is required for all entries; each row must always have a nonnull value for the intersecting field. The superkey for this table is (lang-code), as each language is unique.

Storing data corresponds to inserting new rows or entities into a table. For example, to add another language, Italian, to this table, we must enter the language code (IT) and the name (Italian).

Two tables can be related through matching values. For instance, the tables FOOD and FOOD-LANGUAGE are related by the key food-code, the tables LANGUAGE and FOOD-LANGUAGE by the key lang-code. As a result, an entry in the FOOD-LANGUAGE table is meaningful only if its keys match existing ones in the FOOD and LANGUAGE tables. Because value-based relationships are so essential to a relational data base, it is important to avoid dangling references. That is, one must be sure that no new food name is added unless its food-code and lang-code values match existing codes in the FOOD and LANGUAGE tables. Likewise, no food-code or lang-code value should be changed or deleted in the FOOD or LANGUAGE tables if these codes are used in the FOOD-LANGUAGE table. SQL may automatically handle all these constraints.

SQL can also be used to retrieve data. The major statement used is in this case select ... from ... where. Select specifies the data that you want to retrieve from a data base. 'From' specifies the source table or tables for the result table. 'Where' states the predicate for including a row from the source tables in the result table; it is a boolean expression that usually contains reference to a table's columns.

The result of an SQL query is a relation. Let us consider a very simple query, again using an example illustrated in figure 4, 'Give the French name for food code 12005':

```
SELECT food-name
FROM FOOD-LANGUAGE
WHERE lang-code = 'FR' AND food-code = 12005
```

Expert Systems

Information modeling is also very useful for another kind of computer application called expert systems. From a practical point of view, an expert system consists of four parts:

(1) The description of the problem to be analyzed – called the data base; this represents the question asked by the nonexpert user.

(2) A set of logical rules – the knowledge base – applicable to the general context of the expert system; it is built by experts in the specific area.

(3) A program that controls, interprets, and organizes the logical rules; many commercial computer programs perform this task.

(4) An interface to input or display the information and the results of the processing.

An expert system can be efficient when experts are present in the studied area, the amount of information used by the experts is great, and the problem cannot be solved by simple algebraic formulas or algorithms but rather by a complex sequence of logical decisions [53]. These three conditions are met when considering food naming and labeling, in particular.

Thus, an expert system was developed to help food manufacturers in naming and labeling foods. It works from data input by the user concerning food ingredients and required composition. The regulatory references are retrieved by backward chaining, and a regulatory label is proposed [54].

Such a program with an expert system could follow this scenario: the user first inputs the description of the food product he wants to name by answering a list of questions concerning, for example, the pasteurization, the antioxidants added, or the origin of the raw material. If the user does not know the answer or needs more information, he can input a blank and the expert system asks complementary questions. Regulatory information is displayed, such as a list of permitted antioxidants, as well as legislative text references and summaries. After all questions have been answered, one or several food names are proposed based on the regulations. The regulatory food name can then be used as a key to develop the label based on regulatory food label requirements stored in a data base.

Concluding Remarks

The application of the entity-relationship model of information to food composition that is described here can be compared to a patchwork quilt. We have presented the basic knowledge that supports food composition data to store and certify nutritional data, that is, description of foods and constituents, computation formulas, regulations, etc. Many other entity sets could be added, depending on the context and needs of the data bank. Thus, entity sets are like patches and relationships like connecting thread. One patch can be removed or added; what remains still works independently. This is a great advantage for program development and data base evolution.

Of course, data modeling is not the final step. The next step is software development, which is only superficially described here because it interests computer people more than food scientists. However, programming requirements may sometimes entail slight modifications of the model, according to software capacities.

The scientific definition of foods, beyond a purely technical use, represents economical – and soon legal – stakes, in the framework of new worldwide food trade. In fact, one of the first aims of a data bank containing updated food composition is to homogeneously structure information coming from different laboratories. Each chemical result in the data bank must specify its origin (laboratory or literature), the exact nature of the food, the precise name of the constituent measured, the unit of measure, and the analytical method used. This standardization of vocabulary is accomplished by appropriate coding systems.

Validation of analytical data is correlated to the validation of food information. The real challenge remains validation of nutrient information. The precision of analytical methods in food chemistry is still very poor, and laboratories need to take part in collaborative studies to reach a level of precision in the food industry comparable to that in other branches of industry. It is clear that data banks are tools in this scientific and economic goal.

References

1 Harris LE, Jager F, Leche TF, et al: International feed names and country feed names. Logan, Utah, International Network of Feed Information Centers (INFIC), 1980.
2 Harris LE: A system for naming and describing feedstuffs, energy terminology, and the use of such information in calculating diets. J Animal Sci 1963;20:535–547.
3 Harris LE, Aspund JM, Crampton EW: An international feed nomenclature and methods for summarizing and using feed data to calculate diets. Utah Agric State Bull 1968;479.
4 International Network of Food Data Systems (INFOODS), Room 20A-226. Massachusetts Institute of Technology, Cambridge, Mass, USA.
5 Rand WM: Food composition data: Problems and plans. J Am Diet Assoc 1985;85: 1081–1083.
6 Journal of Food Composition and Analysis, Dr. Kent K. Stewart (ed), Department of Biochemistry and Nutrition, Virginia Polytechnic Institute and State University, Blacksburg, VA 24061, USA.
7 US Department of Agriculture, Human Nutrition Information Service, Agriculture Handbook No 8 Series: 16 vol (1976–1986).
8 Fourteenth National Nutrient Databank Conference, 19–21 June 1989, University of Iowa, Iowa City, Iowa, USA.
9 Recensement de logiciels de nutrition. Orsay (France), Boutique de Sciences – Université de Paris-Sud, 1987.
10 Holman PCH, Katan MB: Report of the Eurofoods interlaboratory trial 1985 on laboratory procedures as a source of discrepancies between food tables. Wageningen (The Netherlands), State Institute for Quality Control of Agricultural Products, 1985.
11 Commission of the European Communities, Community Bureau of Reference: Applied metrology and chemical analysis. The BCR program 1983–1987: Projects and results. Directorate General for Science, Research and Development.
12 Belliardo JJ, Wagstaffe PJ: BCR reference materials for food and agricultural analysis: An overview. Fresenius Z Anal Chem 1988;332:533–538.
13 Feinberg M, Ireland-Ripert J: La banque de données sur la composition des aliments: Un enjeu économique et technique pour l'industrie agroalimentaire. Bull Acad Vét 1986;59:211–224.
14 Favier JC, Feinberg M: La banque de données sur la composition des aliments: Un atout pour l'industrie laitière. Rev IAA, May 1986, pp 404–410.
15 Bugner E, Chaisemartin D, Favier JC, et al: La banque des données sur la composition des aliments: Pourquoi, pour qui et comment. Réunion de la Société Nutrition Diététique Langue Française, Lille (France) 1988.
16 Feinberg M, Favier JC, Ireland-Ripert J: Répertoire Général des Aliments, vol 1: Les Corps Gras. Paris, Lavoisier Tec et Doc, 1987.
17 Feinberg M, Favier JC, Ireland-Ripert J: Répertoire Général des Aliments, vol 2: Les Produits Laitiers. Paris, Lavoisier Tec et Doc, 1987.
18 Holland B, Unwin ID, Buss DH: Cereals and Cereal Products. Nottingham, Royal Society of Chemistry, 1988.
19 Tan SP, Wenlock RW, Buss DH: Immigrant Foods, London, HMSO, 1985.

20 Feinberg M, Favier JC, Ireland-Ripert J: Le Concept REGAL: Une banque de données sur les aliments, pour quoi faire? Sci Aliment 1987;7:355–360.

21 Soyeux A, Feinberg M, Geslain-Lanéelle C: Elaboration d'un dictionnaire réglementaire des aliments. Colloque AGORAL 26–27 April 1988, pp 270–275.

22 Korth HF, Silberschatz A: Database System Concepts. New York, McGraw-Hill, 1986.

23 Robert & Collins French/English, English/French Dictionary. Paris, Le Robert-Collins, 1984.

24 Reinberg M, Favier JC, Ireland-Ripert J: Problèmes liés à la codification des aliments: Le glossaire descriptif et analytique des aliments. Paris, Congrès Aliment 2000, 1988.

25 CODATA, 51, boulevard de Montmorency, F–75016 Paris, France.

26 Huffenberger MA, Wigington RL: Chemical Abstracts Service approach to management of large databases. J Chem Info Comp Sci 1975;15:43–47.

27 International Directory of Food Composition Tables. Cambridge, Mass, International Network of Food Data Systems, 1987.

28 Arab L, Wittler M, Schettler G: European Food Composition Tables in Translation. Berlin, Springer, 1987.

29 Dr Lenore Arab-Kohlmeier, Department of Epidemiology of Health Risk, Federal Health Office Berlin.

30 Haendler H: Synthetic description systems for accurate data identification and selection: Principles and methods of nutritional data banks. Int Classif 1988;15: 64–68.

31 McCann A, Pennington JAT, Smith EC, et al: FDA's Factored Food Vocabulary for food product description. J Am Diet Assoc 1988;3:336–341.

32 Herold P: Using the Factored Food Vocabulary in Food Composition Databases. National Cancer Institute, Bethesda, 1987.

33 Smith EC: Update on Factored Food Vocabulary: Langual. Fourteenth National Nutrient Databank Conference, 19–21 June 1989, University of Iowa, Iowa City, Iowa, USA.

34 Pennington JA: Total Diet Study and Factored Food Vocabulary: Langual. Fourteenth National Nutrient Databank Conference, 19–21 June 1989, University of Iowa, Iowa City, Iowa, USA.

35 Butrum R, Pennington J: Technology systems used for food composition data bases; in Glaeser PS (ed): Computer Handling and Dissemination of Data. Amsterdam, Elsevier Science Publishers, 1987, p 404 ff.

36 Singer DD: Personal communication. Director, NCI Food Component Research Database, Diet and Cancer Branch, EPN 212D, 9000 Rockville Pike, Bethesda, MD 20892, USA.

37 Polacchi W: Standardized food terminology: An essential element for preparing and using food consumption data on an international basis. Food Nutr Bull 1985;8: 66–68.

38 Klensin JC, Feskanich D, Lin V, et al: Identification of food components for INFOODS data interchange. INFOODS 1988;IS N40.

39 Dehove R: Réglementation des produits, qualité, répression des fraudes. Paris, Lamy, 1986–1989.

40 Code of Federal Regulations. Washington, Office of the Federal Register, National Archives and Records Administration of the United States.
41 US Food and Drug Administration. Factored Food Vocabulary: Scope notes.
42 Bulletin International des Douanes. Bruxelles, Communauté Economique Européenne, 1983.
43 IS 22 memo: Representation of trace, missing, and zero values in food data interchange.
44 Feinberg M: Méthodes statistiques et informatiques de valorisations de données de laboratoire: Application à la conception d'une banque de données sur les aliments. Assoc Chimistes de Ind Agro-Alimentaires (ACIA), 16 October 1986.
45 Feinberg M, Bugner E: Chemometrics and food chemistry: Data validation. Anal Chim Acta 1988;191:75–85.
46 National Institute of Standards and Technology (NIST), Office of Standard Reference Materials, Gaithersburg, MD 20889, USA.
47 Laboratory of the Government Chemist (LGE), Office of Reference Materials, Queen's Road, Teddington, Middlesex TW11 OLY, UK.
48 Bundesanstalt für Materialforschung und -prüfung (BAM), Fachgruppe 1.4, Berlin.
49 Laboratoire National d'Essais (LNE), 1, rue Gaston-Boissier, F–75015 Paris, France.
50 IBM Corporation: SQL/Data Terminal User's Guide, 1982.
51 Oracle Company: Oracle User's Guide, 1983.
52 Data General Corporation: DG/SQL User's Manual, 1986.
53 Taylor W: Introduction to artificial intelligence and its applications. Design News, March 1986, p 75 ff.
54 Geslain-Lanéelle CMA, Soyeux AP, Feinberg MH: Expert system for food labeling. Food Technol 1989;43:100–103.

Max H. Feinberg, PhD, Centre Informatique sur la Qualité des Aliments,
16, rue Claude-Bernard, F–75231 Paris Cedex 05 (France)

Simopoulos AP, Butrum RR (eds): International Food Data Bases and Information Exchange, World Rev Nutr Diet. Basel, Karger, 1992, vol 68, pp 94–103

LanguaL

An Automated Method for Describing, Capturing and Retrieving Data about Food

Thomas C. Hendricks

US Food an Drug Administration, Center for Food Safety and Applied Nutrition, Washington, D.C., USA

Contents

Introduction

LanguaL, which stands for '*Langu*e des *aL*iments, or language of food', is an automated method for describing, capturing and retrieving data about food, It has been developed by the Center for Food Safety and Applied Nutrition (CFSAN) of the US Food and Drug Administration (FDA) over the last 15 years as an ongoing cooperative effort of specialists in food technology, information science and nutrition.

LanguaL is based on the concept that: (1) any food (or food product) can be systematically described by a combination of characteristics; (2) these characteristics can be categorized into viewpoints and coded for computer processing, and (3) the resulting viewpoint/characteristic codes can be used to retrieve data about the food from external data bases.

This paper will describe LanguaL in some detail. It will then present an in-depth example of how LanguaL can be applied. Finally, it will review LanguaL's use internationally and postulate its acceptance as a worldwide standard.

How Is Food Described?

One can systematically describe any food or food product using carefully selected points of view, in order to categorize its: (1) food group; (2) origin; (3) physical attributes; (4) processing; (5) packaging; (6) dietary uses, and (7) miscellaneous characteristics.

Each viewpoint is called a 'factor'. There are presently 13 factors in LanguaL (as implemented in the USA). The characteristics related to them are called 'factor terms'. Table 1 presents each factor by category together with a brief description (where appropriate) and examples of factor terms.

As an example, consider the commercial food product 'Corn Flakes' (fig. 1). Its *product type* factor term, which identifies the food group it

HOW IS FOOD DESCRIBED?

Each viewpoint is called a FACTOR.
The characteristics related to it are called FACTOR TERMS.

Consider the food product CORN FLAKES:

FACTOR	FACTOR TERM
PRODUCT TYPE	Breakfast cereal
FOOD SOURCE	Field corn
PART OF PLANT OR ANIMAL	Seed or kernel, skin removed, germ removed (endosperm)
PHYSICAL STATE, SHAPE OR FORM	Whole, shape achieved by forming, thickness < 0.3 cm
EXTENT OF HEAT TREATMENT	Fully cooked
COOKING METHOD	Cooking method not applicable
TREATMENT APPLIED	Sucrose added; flavoring or spice extract or concentrate added; vitamin added; iron added; flaked; water removed
PRESERVATION METHOD	Dehydrated or dried
PACKING MEDIUM	No packing medium used
CONTAINER OR WRAPPING	Paperboard container with paper liner
FOOD CONTACT SURFACE	Wax; paperboard or paper
CONSUMER GROUP/DIETARY USE	Human food, no age specification, regular diet
ADJUNCT CHARACTERISTICS OF FOOD	None

1

Table 1. LanguaL factors and factor term examples

Food group		
A. Product type	Derived from a combination of consumption, functional and manufacturing characteristics	Dairy product, poultry/poultry product, beverage, gravy or sauce, sweetener
Origin		
B1. Food source	Species of plant or animal, of chemical food source	Cattle, abalone, wheat, carob bean, garlic
B2. Part of plant or animal		Leaf, fruit, skeletal meat, organ meat
Physical attributes		
C. Physical state, shape or form		Liquid, semiliquid, solid, whole natural shape, divided into pieces
Processing		
D1. Extent of heat treatment		Fully cooked, partially cooked, uncooked, raw
D2. Cooking method	Cooked by dry or moist heat; cooked with fat; cooked by microwave	Sautéed, baked or roasted, griddled, toasted, popped, deep-fried
D3. Treatment applied	Additional processing steps, including adding, substituting, or removing components	Enriched, sweetened, egg added, fat or oil added, fat removed, decaffeinated
D4. Preservation method	Primary preservation method	Dehydrated or dried, frozen, preserved by adding chemicals
Packaging		
E1. Packing medium		Packed in broth, packed in gelatin, packed in gravy or sauce
E2. Container or wrapping	Container material, form, and possibly other characteristics	Paperboard tray with wrapper, plastic boil-in-bag, glass container, aluminum lid, plastic lining
E3. Food contact surface	The surface(s) with which the food is in contact	Ceramic, cork, paperboard, glass, metal, plastic
Dietary uses		
F. Consumer group/ dietary use	Human or animal; special dietary characteristics	Human food low fat, human food sodium-free, human food very low sodium, human food reduced calorie
Miscellaneous characteristics		
Z. Adjunct characteristics of food	Additional miscellaneous descriptors	Pink fish flesh, shoulder (meat cut), choice (grade), edible sausage casing, mold rind, dry mix

belongs to based on common consumption, functional or manufacturing characteristics, is 'breakfast cereal'. The *food source* factor term, which identifies the individual plant or animal from which the food product (or its major ingredient) is derived, would be 'field corn'. The *part of plant or animal* factor term, which describes the anatomical part of the plant or animal from which the food product (or its major ingredient) is derived, is 'seed or kernel, skin removed, germ removed (endosperm)'. Its *physical state, shape or form* factor term, which distinguishes between liquids and solids and further describes solids in terms of shape or form, is 'whole, shape achieved by forming, thickness less than 0.3 cm'. The *extent of heat treatment* factor term, which is used to broadly characterize a food product based on the amount of heat applied, is 'fully cooked'. The *cooking method* factor term, which identifies the way food products are cooked, reheated or warmed, is 'not applicable' because heat treatment is inherent in the process of making Corn Flakes. The *treatment applied* factor term, which is used to describe components added or subtracted as well as processes involved in producing the food product and can have multiple values, includes: (1) sucrose added; (2) flavoring or spice extract or concentrate added; (3) vitamin added; (4) iron added; (5) flaked, and (6) water removed.

The *preservation method* factor term, which identifies the primary method for preventing or retarding microbial or enzymatic spoilage of a food product, is 'dehydrated or dried'. The *packing medium* factor term, which describes the medium in which the food is packed for preservation, handling and/or palatability, is 'no packing medium used' in this example. The *container or wrapping* factor term, which identifies the material comprising the main container as well as any liners, lids or ends, is 'paperboard container with paper liner'. The *food contact surface* factor term, which identifies the specific container materials in direct contact with the food product and can have multiple values, includes: (1) wax and (2) paperboard or paper.

The *consumer group/dietary use* factor term, which specifies who uses the food product, is 'human food, no age specification, regular diet'. The *adjunct characteristics of food* factor term, which allows for miscellaneous descriptions, does not have any value for Corn Flakes.

As another example, consider the homemade food product 'Breaded, Fried Chicken'. Figure 2 displays its factors and factor terms. As can be seen from these examples, virtually any food or food product can be categorized and characterized in this systematic fashion.

HOW IS FOOD DESCRIBED?

Consider the food product
HOMEMADE BREADED
FRIED CHICKEN:

FACTOR	FACTOR TERM
PRODUCT TYPE	Poultry or poultry product
FOOD SOURCE	Chicken
PART OF PLANT OR ANIMAL	Skeletal meat part, with bone, with skin
PHYSICAL STATE, SHAPE OR FORM	Whole, natural shape
EXTENT OF HEAT TREATMENT	Fully cooked
COOKING METHOD	Cooked with added fat
TREATMENT APPLIED	Fat or oil coated; Grain added; Breaded, batter-coated
PRESERVATION METHOD	No preservation method used
PACKING MEDIUM	No packing medium used
CONTAINER OR WRAPPING	No container or wrapping used
FOOD CONTACT SURFACE	No food contact surface present
CONSUMER GROUP/DIETARY USE	Human food, no age specification, regular diet
ADJUNCT CHARACTERISTICS OF FOOD	Meat color, mixture

2

How Are Factor Terms Identified?

In order to facilitate computer processing, each factor term is assigned an alphanumeric code identifying the factor and the unique characteristic (e.g., A258 = product type 'breakfast cereal'; A273 = product type 'poultry or poultry product'). Figures 3 and 4 show this coding schema for the Corn Flakes and Fried Chicken examples described above.

How Is Information Stored?

A computer record contains the food (product) name and assigned factor term codes, as well as product identification codes (e.g., Universal Product Code or UPC) for accessing external data bases. Figures 5 and 6 depict these data for the Corn Flakes and Fried Chicken examples described above.

Thus, if the Corn Flakes record were selected as part of a search, it would 'point' the user to additional information in the following data

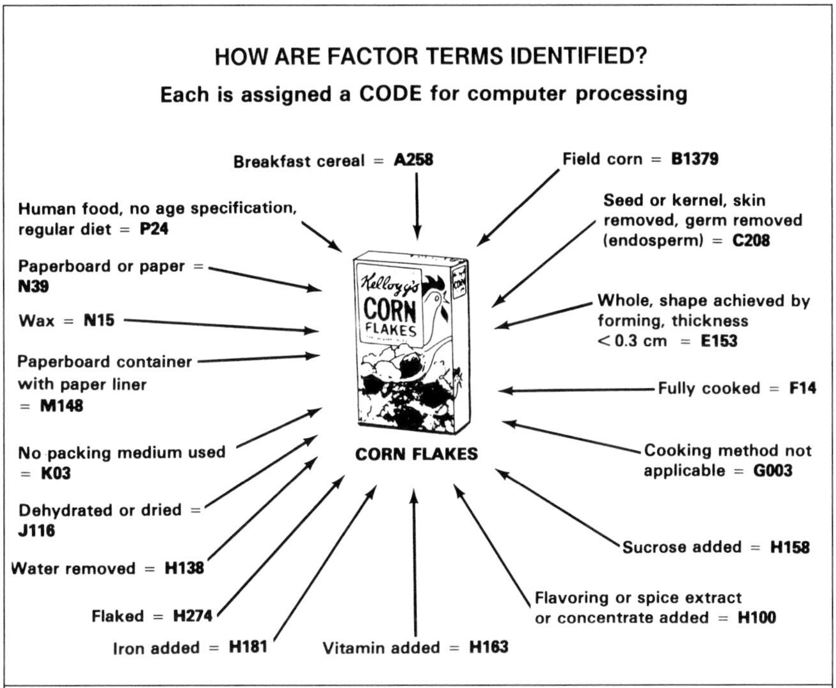

HOW ARE FACTOR TERMS IDENTIFIED?

Each is assigned a CODE for computer processing

Breakfast cereal = **A258**

Field corn = **B1379**

Human food, no age specification, regular diet = **P24**

Seed or kernel, skin removed, germ removed (endosperm) = **C208**

Paperboard or paper = **N39**

Wax = **N15**

Whole, shape achieved by forming, thickness < 0.3 cm = **E153**

Paperboard container with paper liner = **M148**

Fully cooked = **F14**

No packing medium used = **K03**

CORN FLAKES

Cooking method not applicable = **G003**

Dehydrated or dried = **J116**

Water removed = **H138**

Sucrose added = **H158**

Flaked = **H274**

Flavoring or spice extract or concentrate added = **H100**

Iron added = **H181**

Vitamin added = **H163**

3

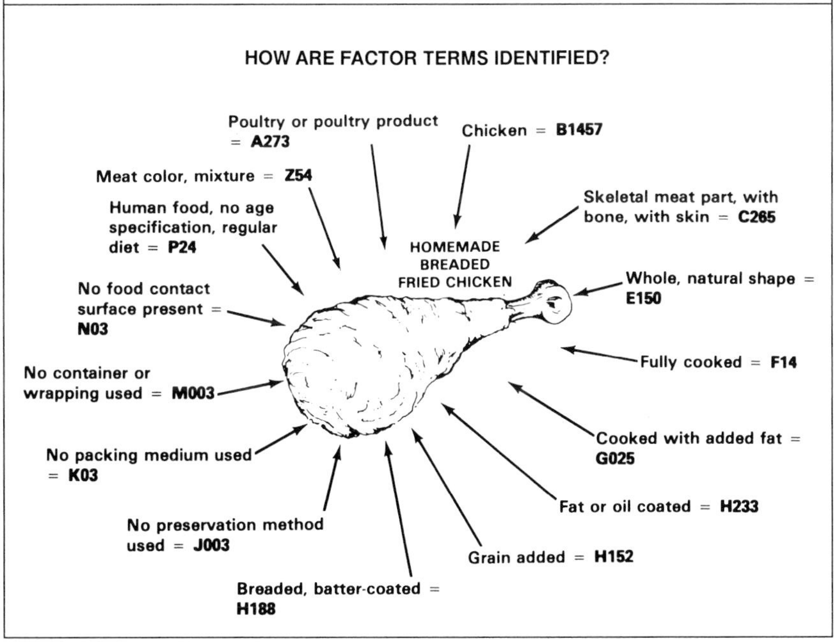

HOW ARE FACTOR TERMS IDENTIFIED?

Poultry or poultry product = **A273**

Chicken = **B1457**

Meat color, mixture = **Z54**

Human food, no age specification, regular diet = **P24**

Skeletal meat part, with bone, with skin = **C265**

HOMEMADE BREADED FRIED CHICKEN

No food contact surface present = **N03**

Whole, natural shape = **E150**

No container or wrapping used = **M003**

Fully cooked = **F14**

No packing medium used = **K03**

Cooked with added fat = **G025**

Fat or oil coated = **H233**

No preservation method used = **J003**

Grain added = **H152**

Breaded, batter-coated = **H188**

4

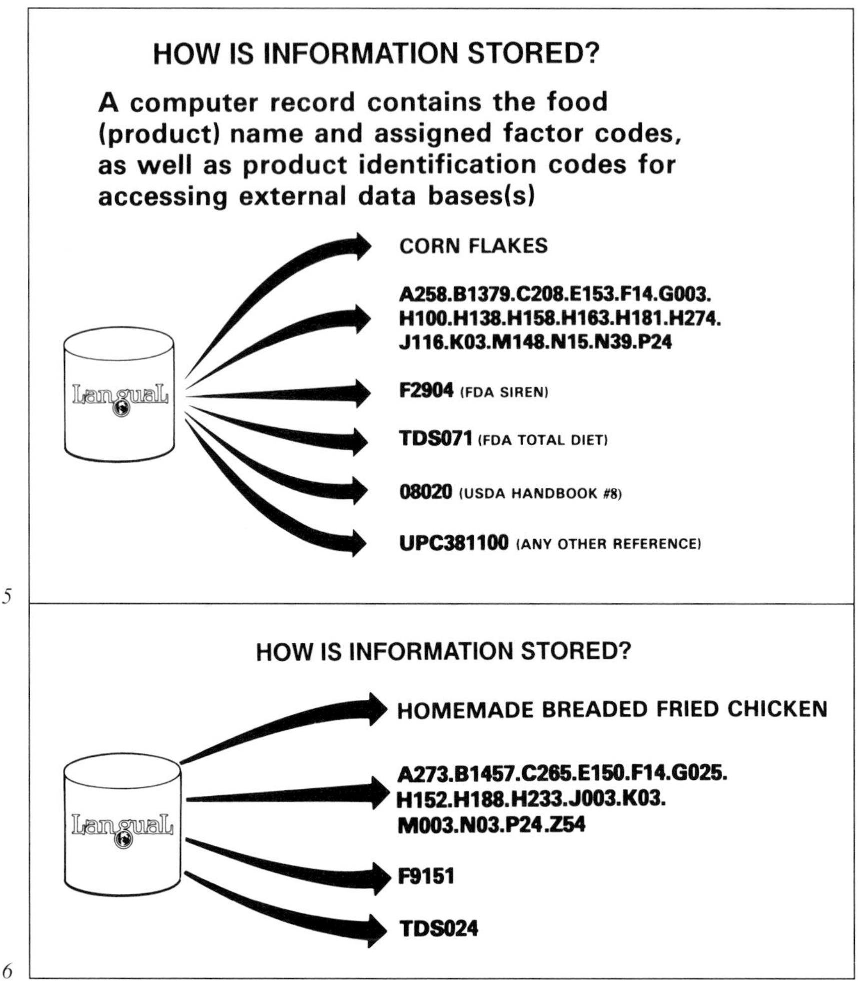

5

6

bases: (1) FDA SIREN[1]; (2) FDA total diet[2], and (3) USDA Nutrient Data Base (Handbook No. 8)[3], as well as any other references (e.g., UPC, where applicable). The Fried Chicken record would have similar 'pointers' to additional data. This means that a wide variety of food composition and

[1] *S*cientific *I*nformation *R*etrieval and *E*xchange *N*etwork, FDA/CFSAN.

[2] Total Diet Study, FDA/CFSAN.

[3] Nutrient Data Base for Standard Reference, US Department of Agriculture (USDA).

food consumption data bases can be referenced by a single LanguaL search, using their existing coding schemes without change. Concomitantly, there is no need to know the codes or structure of the data bases being simultaneously searched by LanguaL.

How Is Information Retrieved?

In general, a LanguaL search is conducted to answer a specific question. The records selected as a result of this search contain pointers to outside data bases. These data bases may then be accessed /using the pointers) to obtain very detailed information related to the original query.

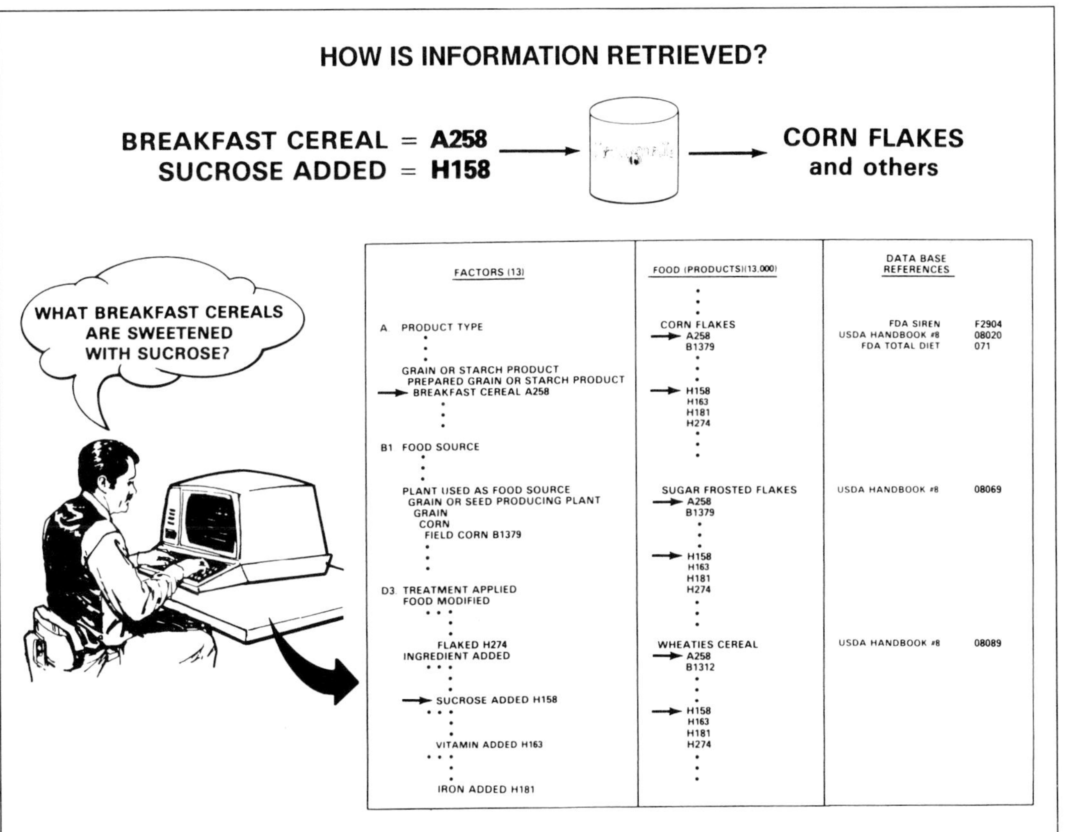

7

For example, consider the question: 'What breakfast cereals are sweetened with sucrose'? By conducting a Boolean search of all the coded foods for those whose LanguaL *product type* factor is 'breakfast cereal' (A258) *and* whose *treatment applied* factor (multiple occurrence) includes 'sucrose added', a number of food (product) records will be selected, including the Corn Flakes example described above. Figures 7 and 8 depict this process. Each selected record contains at least one pointer to an external data base where additional information related to the original query can be obtained. Perhaps data about regulatory (compliance) activities related to sucrose-sweetened breakfast cereal from FDA's SIREN data base might be of interest. On the other hand, food composition and nutrient data for sucrose-sweetened breakfast cereals from USDA's Nutrient Data Base (Handbook No. 8) could be obtained. Moreover, information on residues from con-

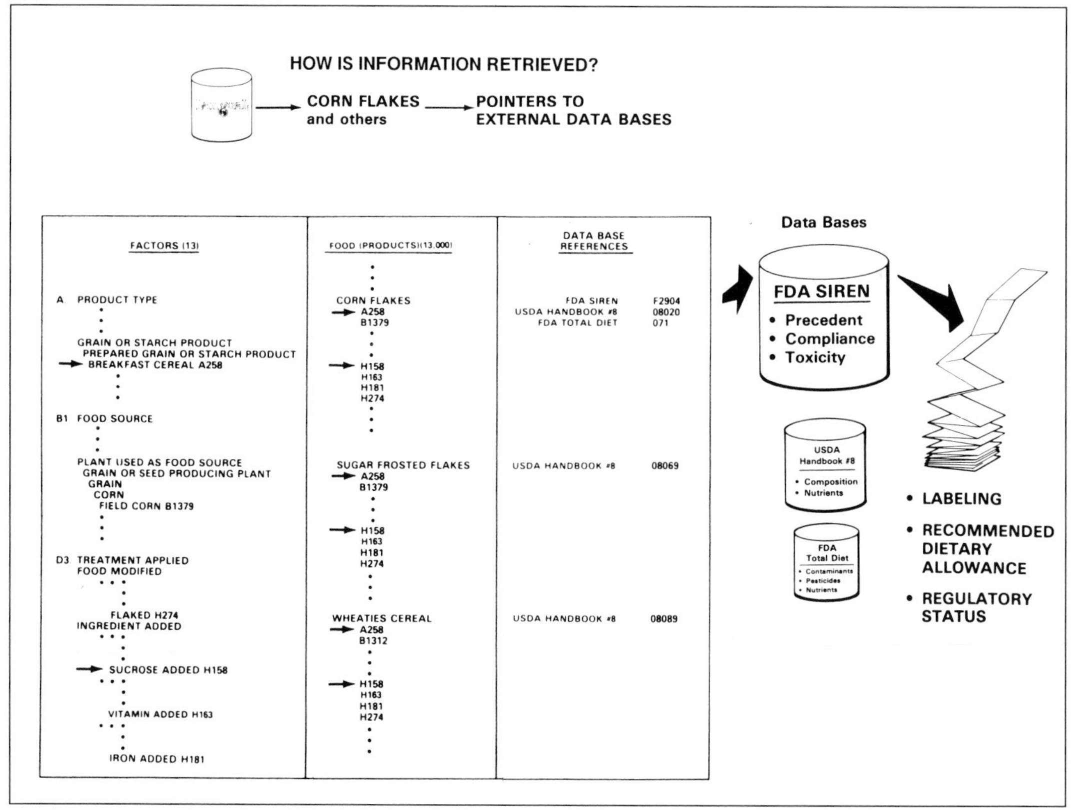

8

taminants and pesticides as well as additional nutritional data could be readily accessed in FDA's Total Diet Study data base. Thus, this one LanguaL query would turn up a wealth of information about sucrose-sweetened breakfast cereals. Such information would be difficult (if not impossible) to obtain in any other way.

Where Does This Lead?

LanguaL can facilitate direct links to many different food consumption and analytical data bases as well as bibliographic files, worldwide. In addition to the FDA and USDA linkages discussed above, LanguaL is also being used by the National Cancer Institute (NCI) in its studies of diet's relationship to cancer incidence. Furthermore, LanguaL is being applied on an international level, with government agencies in Denmark, France and the United Kingdom using it for their endeavors.

At FDA/CFSAN a conscientious effort is being made to expand the number and variety of outside data bases accessible using LanguaL. For example, all 1988 USDA Nationwide Food Consumption Survey (NFCS) foods have been incorporated, thus providing a valuable link to up-to-date food consumption data for the USA. Also, foods and food products whose standards are specified by Title 21 (Foods and Drugs) of the Code of Federal Regulations (CFR) have been coded. This allows access to any data base with CFR food references.

In addition, all Codex Alimentarius food standards (analogous to CFR standards discussed above) have been coded so that any data base with Codex references can be easily accessed. This should prove useful in international trade.

With the advent of the European Economic Community (EEC) in the early 1990s, there will be a need for standardized food descriptions as an adjunct to 'harmonized' tariff schedules. LanguaL could prove very useful in satisfying this need. In fact, it may provide the only effective way for linking with data bases from the Eastern Bloc nations as they become a force in international commerce. With more widespread use, LanguaL could well become an international standard.

Thomas C. Hendricks, US Food an Drug Administration,
Center for Food Safety and Applied Nutrition, 200 C Street SW, Room 6820,
HFF 8, Washington, DC 20204 (USA)

Simopoulos AP, Butrum RR (eds): International Food Data Bases and Information Exchange. World Rev Nutr Diet. Basel, Karger, 1992, vol 68, pp 104–120

NORFOODS Computer Group[1]

Food Composition Data Interchange among the Nordic Countries: A Report

Anders Møller

The National Food Agency, Søborg, Denmark

Contents

[1] The members of the NORFOODS computer group are: Anders Møller, Denmark, Kimmo Louekari, Finland; Trond Ydersbond, Norway, and Hernan Isakson, Sweden.

Introduction

At the request of the Nordic group for Diet and Nutrition Questions, NORFOODS started its work in 1982. NORFOODS is the Nordic group of projects concerning food composition tables and data banks. Since 1982, the members of the NORFOODS group have met each year to discuss and solve problems in relation to collection of data for food composition data banks and the preparation of food composition tables.

The NORFOODS group consists of a representative from each Nordic country: Denmark, Finland, Iceland, Norway, and Sweden. The members of the group are: Anders Møller, Denmark; Maarit Ahola, Finland; Olafur Reykdal, Iceland; Arnhild Haga-Rimestad, Norway, and Lena Bergström, Sweden.

In the autumn of 1985, NORFOODS formed a special working group with the main task of appraising methods for simple transfers of food data by using existing systems of telecommunication or (machine-readable) media. Examination of the technical problems arising from this work was made at two meetings in connection with ordinary NORFOODS meetings held in Copenhagen (December 1985) and in Helsinki (May 1986). The actual data transfer was carried out between and after the two meetings. The final evaluation was made at the end of 1986 and the beginning of 1987. The special electronic data processing group consisted of: Anders Møller, Denmark; Kimmo Louekari, Finland; Trond Ydersbond, Norway, and Hernan Isakson, Sweden.

This report is the result of the group's evaluations. As far as the group knows, it describes the first systematic interchange of data of food composition tables in the world. Furthermore, it suggests a minimum standard run on data format by electronic data transfer of food data among the Nordic countries. It seems to us that this standard could be applied for data transfer between most other countries that have developed food composition data banks.

Background of the Project

All Nordic countries have official food composition tables. The development within the field of electronic data processing, together with the ever-increasing flow of new data of analysis and mutual use of data for equal foods in the Nordic countries, has made possible the better coordination and interchange of data either by telecommunication network or (machine-readable) media. The use of dietetic software in microcomputers has increased the demand for food composition data. Expression of this is also made by the comprehensiveness of the food composition tables from the different countries.

Originally, the basis of food composition tables and dietary calculations is a type of food composition data bank stored in a mainframe computer. With the development of electronics, these tasks have been taken over by the microcomputers, which are easier to use. The construction and maintenance of such data banks demand many resources. Some of the workload could be saved if food data were transferred among the Nordic countries in an electronically readable form.

With respect to these obvious cooperation possibilities, the group of electronic data processing in NORFOODS has evaluated the computer facilities in the individual countries to find common communication contacts. This report has been compiled in an effort to ease the work involved in the future exchange of food composition data. The developments in the field of software for food composition data banks and dietary calculations have also initiated a discussion about possible future areas of cooperation within this field.

Hardware in the Individual Countries

Denmark handles the food composition data bank on an IBM-compatible microcomputer (a PC). The microcomputer has the possibility of communication with the minicomputer, type PDP-11/44, at the National Food Agency, which is supplied with a tape station and with public communication networks by modem.

Finland is currently setting up a food composition data bank for public management. A mainframe-based food composition data bank is available at the Department of Nutrition at the University of Helsinki in Viikki, and the Finnish Pension Institute has a more extended mainframe system called NUTRICIA. A microcomputer is used in the daily work. Some microcomputer applications were developed after 1987. The updating of food composition data banks used by micros has been established in a few cases.

Norway operates the official food composition data bank on an IBM-compatible microcomputer with the possibility of communication by modem with the public communication networks.

Sweden operates the official Swedish food composition table using a system that contains software for the calculation of food, KOST, on the minicomputer NORD-100 with a tape station at the National Food Administration, Uppsala. It is possible to communicate directly with public communication networks and microcomputers.

Iceland has its own food composition data bank on an IBM-compatible microcomputer, along with the possibility of communication by modem with the public communication networks.

Possibilities of Data Interchange

In the table shown below, the direct communication possibilities of the individual Nordic countries are indicated in relation to the individual food composition data bank:

Country	Public communication[1]	Magnetic tape	Diskette
Denmark	(+)	+	+
Finland	(+)	+	+
Norway	(+)	+	+
Iceland	(+)	–	+
Sweden	+	+	+

[1] The completely usable facilities are marked with +; facilities only used from the country named to another country with completely usable facilities are marked with (+).

Public Communication

Mainframe systems have possibilities for data transmission by data network connections, for example, the EARN network, which at the time of the study could be used for research purposes (paid by IBM until end of 1987). Denmark used the EARN network intensively at the end of 1986 and the beginning of 1987 to communicate with the secretariat of INFOODS at the Massachusetts Institute of Technology in Boston, Mass. (described later). However, the problem with the network solution is that it may be expensive for the individual user if the institute concerned is not a classified EARN user.

If you estimate the actual requirement for a physical network connection among the food composition data banks, the conclusion is that it is not needed in the future. It might be part of a possible network system among the Nordic food administrations.

The need is very small for food composition data banks because the amount of data transferred among the individual countries is limited. Typically a food composition data bank occupies 0.3–2 MB, which means that relatively simple data transmission facilities are sufficient.

Dial-Up Telephone Connections

Dial-up modem connections are also reliable, but they demand a lot of standardization work, for example, according to emulation of terminals. Disturbance on such lines also may be found, which does not make satisfactory data transmission possible. This type of data transmission has been tested from the different Nordic countries to the data centers where dial-up connections are possible.

The connection between Copenhagen and the National Food Administration in Uppsala to the KOST system was tested. There were no problems with the connection itself, but it was not possible to get access to the system with the first try, as the Danish terminal (IBM PC/XC with ordinary asynchronous TTY-emulation) could not emulate the terminal the SLV system demanded. The connection between Copenhagen and the University of Helsinki was tested without problems. There was, however, a bad telephone connection, which was disconnected during the session. The connection from Copenhagen to the University of Oslo took place without problems. The connection from Uppsala to UNI-C in Copenhagen also took place without problems as did the connection from Helsinki to UNI-C with file transmission (information on nutrient losses during preparation of foods).

These very small tests have shown some of the other problems concerning direct connection between the individual countries in addition to those of delivering passwords or account numbers. Finland, Norway, Sweden, and Iceland use the modem connection of universities' centers and not their own system. Permission to connect to the minicomputer system in the National Food Agency in Denmark could solve the problems. The latest development had made possible a direct connection between PCs, if the involved PCs are supplied with newer modems following HAYES/XMODEM standard.

Exchange of Data by Diskettes and Magnetic Tapes

The last interchange possibility of machine-readable data is the direct interchange of data on diskette or magnetic tape. This interchange of data is practicable as all countries have direct access to microcomputer equipment compatible with the equipment in the other Nordic countries. Furthermore, it is the easiest interchange method in the first stage. The interchange of data by diskettes implies other problems too, such as which data format to choose. The term data format means how the data are represented in the data material and in which unit.

This is a problem that appears with all forms of data interchange. Apparently, there were no mechanical problems during data interchange by diskettes. Therefore, the computer group found it suitable to get experience through data interchange by diskettes in the first stage. The goals were first to get experience with converting foreign data to one's own format and then to start the data interchange at an early stage.

Data files containing food composition data bank values were interchanged among all Nordic countries. There were very few limitations to the format used, for example, diskettes containing the data files should be formatted under MS-DOS (PC-DOS), and the data files should contain two files in ASCII form (standard text format). It should be emphasized that there should be no other limitations on the data format because our purpose was to test whether a free data format is subject to any problems for the receiver of the data. The experiences from the data interchange trial were used in the follow-up discussions about common data format.

Denmark, Finland, Iceland, and Norway sent diskettes to the other Nordic countries; Sweden, in the first stage, sent magnetic tapes to Norway for conversion to diskettes because the National Food Administration in Sweden at the start of the project could not deliver diskettes. It was not possible to convert the Swedish tapes in Norway. Another tape was sent to Denmark, where the conversion eventually succeeded after solving some problems. In the summer of 1987, Denmark received a diskette from the National Food Administration with data from the 1986 version of the Swedish food composition tables. The original magnetic tape contained data from the version from 1978.

The following section describes the data files from the individual countries along with the problems that arose during deciphering of the data files. To make the evaluation as realistic as possible, all the data are transferred to the same system, the Danish food composition data bank, which now can reference the data from the other Nordic countries.

Comparison of the Interchanged Data Files

The following presents a complete description of the interchanged files with general comments that the comparison may bring about. The general description of file material is shown in the table. Two columns appear for Sweden, one for the magnetic tape version and one for the diskette version:

	Denmark	Finland	Iceland	Norway	Sweden	
Datamedia						
Diskette	*	*	*	*	*	
Tape					*	
Documentation						
Paper					*	
In data file		*				
In special file	*		*		*	
Data file						
One	*	*	*	*		
More					*	*
Text format						
7-bit ASCII					*	*
8-bit ASCII	*	*	*	*		
Data representation						
Decimal	*		*	*	*	
Other		*				*
Special flags						
Numeric	*		*			
Alphanumeric				*	*	
Consistency						
Whole data file	*	*	*		*	
Individual deviance				*	*	
Irrelevant information in data file		*	*			

The following paragraphs describe the individual points. An additional description of the data files is made in Supplement A.

Datamedia states machine-readable media for transferring data. Evaluation of the two medias, diskette and magnetic tape, are made. Diskettes are recommended, as only Denmark and Sweden have direct entrance to tape stations, and input from magnetic tape is more difficult to handle than from diskettes. Input of data from magnetic tapes often requires the involvement of technical personnel while the receiver can make input from diskettes himself. Also, magnetic tape stations usually are not available for the microcomputers.

Documentation of the data material is a necessity for data to be interpreted from the data file. There are considerable difficulties with the deciphering if the documentation is stored in the same file as the data. Elimination of the documentation from the data file has to be done before input of data can take place. The documentation must be placed alone in a separate file or, if this is not possible, on paper. The documentation must *not* be placed in the data file. This was the case in the Finnish file and partly in the Icelandic.

With the increasing amount of data files, difficulties with data input arise. Included in the term data files are only files containing data on substances and not files with data on recipes or food names. Only Sweden uses several data files, two data files in magnetic tape form (nutrients, fatty acids) and nine data files in diskette form (one for each food group). To facilitate data input, only *one* data file should be used.

The text format states how the data are stored on the diskette. Before the data interchange, it was agreed that the data format should be ASCII standard text format. This is respected by all the involved countries. However, some problems appeared. In the Swedish diskette version, a 7-bit ASCII code with parity check was used. This was not expected; thus some problems appeared during input because the files had to be converted to 7-bit ASCII format without parity check before input. As expected, there were problems with the special national characters (æ, ø, å, ä, ö), as there was no room for these symbols in the original 7-bit ASCII table. Recommendations for using the expanded 8-bit ASCII table are given as far as possible (the IBM PC character set).

The value format is the way the values are stored in the data files. In the Swedish magnetic tape version and in the Finnish data file, the values are not represented by decimal values but are instead represented by integers. The amount of decimals in these values can be found in the documentation files. This causes problems in the data input because conversion of values also has to take place at the data input. Therefore, it is recommended that decimal notation be used.

The special flag is a way to mark special values in the data file. Most often, it will signal missing values or traces, otherwise, an expected value of zero is assumed. In the Danish data file, '−1' is used to mark a 'missing value'. In both the Swedish magnetic tape version and the diskette version the missing value is stated with an empty space, whereas traces in the diskette version are stated with the word 'spår'. In the Norwegian data file, '...' is the missing value. All '–' (almost zero) in the printed Norwegian tables are stated with 0 (zero) in the data file. The Finnish data file does not state any special flags, but it is assumed that some of the zeros in the data file are special flags. The reason for this is the problems that arise when data are converted for nutrient calculation purposes.

These problems were the first to be treated by the project group after the first interchanges of data. It was decided that the same negative values be used to flag these special cases. It was decided that the value '−1' be used to state 'trace' and the value '−9' be used to show 'missing value'. Since the last interchanges of data, there can be a further use of flags, for example, to state 'probably zero value, but not analyzed' (the Norwegian '–') and 'under limit of detection'.

Consistency shows in the total data file whether the same fields are used for the same substances or if a certain field can contain information about different substances. This is the case in the Norwegian data file as well as in the Swedish. In the Norwegian file the alcohol values use the same field as vitamin D values. In the Swedish files alcohol and cholesterol values are found in the same field. As there was no documentation for the Norwegian file, this inconsistency was found only after data input and following the cross-reference to the Norwegian table. It was documented in the Swedish data file. Inconsistency is very inexpedient during data input; therefore, one must strive for each field in a data file to have only *one* meaning for all foods in the file.

Irrelevant data file information appeared in the Finnish and the Icelandic data file, as the first lines in the data file have no absolute meaning for information concerning the rest of the file. They only serve as headings. Such information should be avoided in the data file. It belongs to the documentation file.

The above paragraphs outline the most important comments to the problems that have arisen during the input and conversion of data. It should be noted that there were no problems too big to solve. The interchange would be easier if the data files were more alike.

Proposals for a Minimum Standard for Food Data Interchange

From the tests described herein, it is clear that standardization is needed with respect to the future interchange of data on food composition tables among the different national centers in the Nordic countries.

The minimum claims that the project group finds necessary are described below. The aim is to get an easy and uncomplicated food composition data interchange among the Nordic countries. The following minimum demands have to be met by transformation of food data on machine-readable media:

Data media:
The data media have to be diskettes formatted under MS-DOS (PC-DOS), 5¼ or 3½. In an emergency, a magnetic tape can be used (1,600 BPI, no label).

Documentation:
The documentation has to follow the data media either as a file on the data media or as a separate paper printout.

Data file:
All data should be stored in one data file, eventually including standard file archiving software (this must appear in the documentation).

Text format:
The data format must be 8-bit ASCII standard text format, subsidiary with standard 7-bit text format without parity.

Value format:
The data values must be in decimal format as in the matching food composition table, so that the values are easily recognizable.

Special flags:
Special numeric codes are used to signal special values:
- −1: trace
- −2: value below detection limit (can be fixed to one-third of the level of the detection limit).
- −3: value probably 0 (zero)
- −9: missing information

Consistency:
Every field in a food record must be unambiguous, that is, the same field cannot have different meanings in different food records.

If these minimal demands are kept by food data interchange, it will be possible to incorporate other countries' data in the respective country's own food composition data bank without any other difficulties.

According to the described Nordic food data interchange, the Danish food composition data bank is now extended with reference keys to the other Nordic food composition data banks as well as the British, Dutch, and US food composition data banks.

International Cooperation

Besides NORFOODS, work is carried out concerning food data interchange within the international framework. INFOODS (International Network of Food Data Systems) is operating in the UN framework and is developing a considerably more complex and comprehensive system used to describe food ingredients in relation to food data interchange. The member countries of NORFOODS are also the first countries to take part in the testing of this vast data system specification.

The description of the specification made by INFOODS 'Identification of Food Components for Infoods Data Interchange' can be found in an unpublished paper from the Secretariat of INFOODS, Massachusetts Institute of Technology, Room 20A-226, 77 Massachusetts Avenue, Cambridge, MA 02139, USA.

Conclusion

According to this first test, the possibilities for the individual Nordic countries were examined to find points of contact for technical electronic data processing for immediate use in the interchange of data among the Nordic countries. The project has shown that, with a minimum of restrictions, it is possible to carry out data interchange, which makes it possible for the receiver to recognize and manage the data.

Data interchange in machine-readable form can imply certain problems concerning copyright, especially as it is only the Swedish and the Danish food composition tables that are entirely official and completely paid for by public funds. Therefore, it has been decided that the individual countries' limitations with respect to the use of interchanged data by other countries have to be respected. Because of this, no economic interests concerning this format of data testing can be involved.

Furthermore, the project has shown it to be advantageous to this cooperation that very few people are working within the field of food composition data banks in the individual Nordic countries. This also has implications for better cooperation in the field of data according to the models of data in the food composition data banks, interchange of computing algorithms, interchange and standardization of video screens, and special routines for the import/export of data to/from nutrient calculation software. Also, possibilities of the receiving data system to use the transferred data could be discussed: borrowing of values, when indicated; comparisons and checks with respect to own national data; and identification of sources of food composition data in other data bank-maintaining institutions or in published literature.

Because of the success of the first composition data interchange between the Nordic countries, in November 1989 it was decided that a second data interchange should take place before the end of 1989. Also, it was decided to publish an update report on the food composition data bank systems in the Nordic countries.

Supplement A: Data File Descriptions

A-1

Country of origin: Denmark

Data media: Diskette

Data file layout:

```
ABRIKOS, tørret         24.0           -1
1221                    0              -1
4.8                     5.3            0
0.5                     23.5           0.060
0.2                     617            3.0
0                       0              .
0.3                     3700           .
0                       0              .
65.9                    4.0
41.9                    4.0
```

Remarks: Only one item of information on each line. Numerical special flags (–1 means missing value, according to the proposal in this report the value should be –9).

A-2

Country of origin: Finland

Data media: Diskette

Data file layout:

```
RAVINTOAINETTEDOSTO
NIMET   VALK  RASV HHYDR  KCAL    KJ A-VIT   RET  TIAM   RIB  NIAS C-VIT   KOL 1
NIMET C4-11   C12   C14   C16   C18  MTRH  YTRH  M-ET C18-2 C18-3 MP-ET YP-ET 2
NIMET T#RKK  SAKK  LAKT   MHH   ALK     K    CA    MG     P     S    FE    CU 3
NIMET    MN    ZN     F    SE    MO    BR    RB    AL    SI     B    HG    AS 4
NIMET    CD    CO    CR    NI    PB    NA     I    SR NIASE D-VIT     N  VESI 5
DESIM.    1     1     1     0     0     0     0     2     2     2     1     1 1
DESIM.    2     2     2     2     2     2     2     2     2     2     2     2 2
DESIM.    1     1     1     1     1     4     1     2     1     4     3     3 3
DESIM.    3     3     3     2     3     3     3     3     1     3     1     3 4
DESIM.    1     1     1     3     3     1     3     2     1     2     2     1 5
001933    0     0   999   387  1620     0     0     0     0     0     0     0 1
001933    0     0     0     0     0     0     0     0     0     0     0     0 2
001933    0   999     0     0     0  0020     4    10     1  0100   060   010 3
001933  010   010   050    50   010   100   100   300     5   030     2   003 4
001933    5    10    10     3    10     1     5     1     0     0    10     1 5
002933    0     0    95   380  1589     0     0     0     0     0     0     0 1
002933    0     0     0     0     0     0     0     0     0     0     0     0 2
002933    0     0     0    95     0  0003     1     0     1  0100   070   010 3
002933  010   010   060    50   010   100   100   300     5   030     2   003 4
002933    5    10    10   003   010     1     5     1    00     0     0     1 5
003933   03     0   767   298  1250     0     0     0    01    10     0     0 1
003933    0     0     0     0     0     0     0     0     0     0     0     0 2
003933    0   421     0   346     0  2200   740  3400    20  0300  2500   240 3
003933  110   130   080   100   010   200   290  1600    50   030     2   006 4
.
.
.
```

Remarks: The first line in the data file is irrelevant. The file documentation is on lines 2–11, where as lines 2–6 give information about the substances and lines 7–11 contain information about the decimals in the values stated as integers in the data file.
This information should be included in a special documentation file.
Cleaning of the data file before input must be avoided.

Country of origin: Iceland

Data media: Diskette

Data file layout:

```
FOOD ITEM                   P     F     C     S     F     A     W       B1

Blood sausage, cooked     7.9  22.6  17.5   0.0    -9   1.6  47.9   0.049
Foal goulash             22.8   4.4   0.0   0.0   0.0   1.1  72.3   0.17
Smoked lamb meat         18.0  15.1   0.0   0.0   0.0   4.6  61.3   0.090
Shark                    29.8  47.7   0.0   0.0   0.0   2.1  20.7      -9
Cod roe                  25.0   1.0    -9    -9   0.0    -9  69.8   0.37
Chicken                  19.4  11.9   0.0   0.0   0.0   0.9  69.2   0.10
Lamb leg                 20.2   5.5   0.0   0.0   0.0   1.1  73.8   0.18
Halibut, fillet          19.2   0.9   0.0   0.0   0.0   1.1  79.4   0.07
Lake trout               20.5   2.7   0.0   0.0   0.0   1.2  76.6   0.16
Ham                      18.7   6.9   0.0   0.0   0.0   3.2  71.7      -9
Salami                   15.6  39.4    -9    -9    -9   5.7  36.1    -9
Pollock, fillet          19.3   0.3   0.0   0.0   0.0   1.2  79.5   0.09
Haddock, fillet          18.9   0.2   0.0   0.0   0.0   1.1  80.4   0.03
Wiener sausage           11.2  20.6   5.9    -9    -9   2.8  58.0   0.12
Cod Liver                 5.1  66.6   0.0   0.0   0.0   0.5  26.8   0.08
.
.
.
```

Remarks: The first two lines in the data file should be placed in a special documentation file, as they are irrelevant to the data file.
Uses the special flag '–9' to signal missing values, as proposed by the working group.

Country of origin: Norway

Data media: Diskette

Data file layout:

```
01.001MELK,KEFIR,KULTUR             100 87 281 67 3.3  3.8  4.8  0   0 120 0.1   45 0.04 0.15 0.1   1
01.002LETTMELK                      100 90 198 47 3.4  1.5  4.9  0   0 120 ..    18 0.04 0.15 ..   ..
01.003SKUMMET MELK,SKUMMET KULTUR   100 91 147 35 3.4  0.1  5.0  0   0 120 0.1    0 0.04 0.15 0.1   1
01.004KJERNEMELK                    100 92 168 40 3.4  0.7  5.0  0   0 120 0.1    0 0.04 0.15 0.1   1
01.005KONDENSERT MELK,USUKRET       100 74 588140 7.0  8.0 10.0  0   0 240 0.2  120 0.05 0.35 0.2   1
01.005KONDENSERT MELK,USUKRET       100 74 588140 7.0  8.0 10.0  0   0 240 0.2  120 0.05 0.35 0.2   1
01.007RÅMELK,1.MELKNING             100 74 59614215.0  8.0  2.5  0   0 200 0.1  300 0.16 0.60 0.4   3
01.008SJOKOLADEMELK                 100 82 231 55 3.4  0.5  9.3  0   0 120 0.1    3 0.04 0.15 0.1   1
01.009TØRRMELK AV HELMELK           100  3211750426.0 28.0 37.0  0   0 900 0.7  330 0.30 1.10 0.7   4
01.010TØRRMELK AV SKUMMET MELK      100  4150435835.0  0.7 53.0  0   01300 1.1   12 0.43 1.60 1.1   6
01.011YOUGHURT,NATURELL             100 84 315 75 4.2  3.8  6.0  0   0 150 0.1   42 0.04 0.16 0.1   1
01.012YOUGHURT,M/FRUGT              100 76 420100 3.6  3.3 14.0  0   0 130 0.1   36 0.04 0.15 0.1   1
01.013GETTMELK                      100 88 244 58 3.0  3.2  4.3  0   0 130 0.1   60 0.06 0.02 0.1   1
01.014KVINNEMELK                    100 87 281 67 1.2  3.8  7.0  0   0  33 0.15  51 0.02 0.05 0.2   5
01.015KREMFLØTE,SETERRØMME,35% FETT 100 581407335 2.0 35.0  3.0  0   0  95 0.1  300 0.03 0.10 0.1   0
01.016FLØTE,RØMME,20% FETT          100 74 869207 2.8 20.0  4.0  0   0 100 0.1  180 0.03 0.10 0.1   0
01.017HUSHOLDNINGSFLØTE,14% FETT    100 78 651155 3.0 14.0  4.3  0   0 100 0.1  129 0.03 0.10 0.1   0
01.018FLØTEIS                       100 64 785187 4.3 10.0 20.0  0   0 130 0.1  117 0.05 0.18 0.1   0
01.019FLØTEDESSERTER,FRYSTE         100 66 794189 4.5 11.0 18.0  0   0 150 0.1  120 0.04 0.25 0.1   0
01.020MELKEPUDDINGER                100 75 462110 3.2  3.5 16.5  0   0 110 0.2   39 0.04 0.14 0.1   0
01.021MELKERING                     100 87 281 67 3.3  3.8  4.8  0   0 120 0.1   45 0.04 0.15 0.1   0
01.022RØMMEGRØT,FRYST               100 581193284 5.0 24.0 12.0  0   0 120 0.2  180 0.03 0.10 0.2   0
01.023VANILJESAUS                   100 72 647154 2.7  8.7 16.2  0   0 100 0.1   84 0.03 0.12 0.1   0
01.024BRIE                          100 45155036918.6 32.7    0  0   0 370 1.3  300 0.04 0.30 1.6   0
01.025CAMEMBERT                     100 52126430119.0 25.0    0  0   0 230 1.3  210 0.05 0.41 1.2   0
01.026CHATEAU BLEU                  100 561344320 8.0 32.0    0  0   0 200 0.1  240 0.07 0.46 0.6   0
01.027CHEDDAR                       100 37165539426.6 32.0    0  0   0 740 1.0  230 0.03 0.52 0.1   0
01.028COTTAGE CHEESE                100 79 403 9612.7  4.3  1.5  0   0  90 0.8   54 0.05 0.20 0.1   0
01.029CREME CHATEAU                 100 511491355 9.3 35.0  0.8  0   0 200 0.1  240 0.05 0.24 0.2   0
01.030EDAMER                        100 42148735427.0 27.3    0  0   0 900 0.9  210 0.03 0.37 0.1   0
.
.
.
```

Remarks: Uses '...' to signal traces.

A-5

Country of origin: Sweden

Data media: Magnetic tape

Data file layout:

```
1.001 AVOKADO
1.002 BAMBUSKOTT
1.003 BLEKSELLERI=BLADSELLERI
1.004 BLOMK$L
1.006 BONDB§NOR
1.009 BROCCOLI DJUPFRYST
1.010 BRYSSELK$L=ROSENK$L
1.011 BRYSSELK$L=ROSENK$L DJUPFRYST
1.012 GRODDAR MUNGB§NOR
1.014 B§NOR BRUNA TORKADE
1.015 B§NOR BRUNA KOKTA HERMITISERAD
1.017 B§NOR GR§NA
1.018 B§NOR BR§NA DJUPFRYSTA
1.019 B§NOR GR§NA M LAG HERM
1.020 B§NOOR GR§NA HERM UTAN LAG
1.023 B§NOR VITA TORKADE
1.025 B§NOR VITA I TOMATS$S HERM
1.027 CHAMPINJONER
1.028 CHAMPINJONER M LAG HERM
1.030 CHILIS$S TOMAT
1.031 SALLAD ENDIVE-
1.033 F#NK$L
1.036 GR§NK$L DJUPFRYST
1.037 GR§NSAKSBLANDING DJUPFRYST
1.039 GR§NSAKSSOPPA HERM #TF#RDIG
```

Remarks: File with food names. Sweden is the only country that carries the food names in a special file.
Problems with special Swedish characters in a Danish printer.

A-6

Country of origin: Sweden

Data media: Magnetic tape

Data file layout:

```
M1041 1.004 0653      M104114.486 0065      M1043 1.084 0004
M1041 4.214 0261      M104115.903 0031      M1043 4.214 0391
M1041 4.486 0065      M104116.306 0018      M1043 6.306 0030
M1041 5.903 0031      M104117.400 0014      M1043 7.160 0026
M1041 6.306 0018      M104118.702 0008      M1043 8.702 0008
M1041 7.160 0014      M104119.290 0020      .
M1041 8.702 0008      M1042 1.004 0963      .
M1041 9.290 0020      M1042 8.702 0008      .
M104111.004 0653      M1042 9.290 0029
M104114.214 0261      M1043 1.004 0559
```

Remarks: File with recipes.

A-7

Country of origin: Sweden

Data media: Magnetic tape

Data file layout:

```
1.001 0000300016700074000001200002100016400006300000000000174000029000000        0003000 fortsættes
1.002 000000000240009400000034000170000040000320000000000000600000100000000000020000200 fortsættes
1.003 000011000180009500000090000070000010000360000000000001700003000000000000020000200 fortsættes
1.004 000008000270009100000090000270000020000520000000000003600006000000000000000000030 fortsættes
1.006 000066003550001190000310002500000170001170000000000021000003500000000000030000300 fortsættes
1.009 000000000355008870000080000340000030000540000000000135000022500000000000130001300 fortsættes
1.010 000020000450008520000120000490000040000830000000000033000005500000000000100001000 fortsættes
1.011 000000000360008840000080000330000020000730000000000034200005700000000000100001000 fortsættes
1.012 000000000350008880000060000380000020000660000000000001200000200000000000020000200 fortsættes
1.014 000000003390001120000380002230000150006120000000000001800000300000000000230002300 fortsættes
1.015 000000001090007330000130000440000090002020000000000000600000100000000000080000760 fortsættes
1.017 000012000320009060000070000190000020000710000000000036000006000000000000030000300 fortsættes
1.018 000000000270009160000050000170000010000610000000000031800005300000000000030000300 fortsættes
  .
  .
  .
```

Remarks: File with nutrient values. The format is very hard to recognize.

A-8

Country of origin: Sweden

Data media: Diskette

Data file layout:

```
1.002 BAMBUSKOTT, herm,    a 100   99  24   94    0.3  1.7  0.4  0.1  spår  fortsættes
1.003 BLEKSELLERI =        a 100   77  18   95    0.9  0.7  0.1  spår spår  fortsættes
1.008 BROCCOLI             a  61  150  36   90    0.9  3    0.4  spår spår  fortsættes
1.009 BROCCOLI, fryst      A 100  151  36   91    0.8  3.4  0.3  spår spår  fortsættes
1.006 BONDBØNOR            a  94  315  76   81    1.1  5.6  0.6  0.1  spår  fortsættes
1.007 BONDBØNOR, torkade   A 100 1490  355  11.9  3.1  25   1.7  0.2  0.2   fortsættes
1.014 BRUNA BØNOR,         A 100 1480  350  11.2  3.8  22   1.5  0.4  spår  fortsættes
1.015 BRUNA BØNOR, kokta   A 100  450  108  73    1.3  4.4  0.9  0.2  spår  fortsættes
1.017 GRØNE BØNOR          a  94  152  36   90    0.7  1.82 0.1  spår spår  fortsættes
1.018 GRØNA BØNOR, fryste  a 100  163  39   90    0.5  1.7  0.2  spår spår  fortsættes
1.020 GRØNA BØNOR, herm,   a 100   95  23   93    1.0  1.2  0.1  spår spår  fortsættes
1.074 MUNGBØNOR, torkade   A 100 1490  355  10.7  3.5  24   1.3             fortsættes
1.021 RØDA BØNOR, torkade  a 100 1490  356  10    3.7  22   1.5  0.2  0.1   fortsættes
1.022 RØDA BØNOR,          A 100  520  123  69    1.3  7.8  0.5  spår spår  fortsættes
1.145 VAXBØNOR             a  94  157  38   90    0.7  1.8  0.1  spår spår  fortsættes
1.149 VIGNABØNOR, inkl     A 100 1490  355  10.5  3.5  23   1.5  0.4  0.1   fortsættes
1.150 VIGNABØNOR, svart-   A 100  330  79   80    0.8  5.1  0.3  0.1  spår  fortsættes
1.023 VITA BØNOR, torkade  A 100 1480  355  11    3.9  22   1.6  0.2  0.1   fortsættes
1.024 VITA BØNOR, torkade  A 100  515  123  69    1.4  7.8  0.6  0.2  spår  fortsættes
1.025 VITA BØNOR, herm     B 100  450  108  73    1.8  5.1  1.2  0.2  0.2   fortsættes
1.027 CHAMPINJONER         a  96  130  31   92    0.9  2.1  0.4  spår spår  fortsættes
1.227 CHAMPINJONER,        F 100  100  24   94    0.8  2    0.4  spår spår  fortsættes
1.338 CHAMPINJONER, herm   a 100  126  30   91    1.7  1.9  0.3  spår spår  fortsættes
1.127 CHAMPINJONSOPPA,     a 100  217  52   90    1.2  1    3.7  1    0.7   fortsættes
S1410 CHAMPINJONSOPPA.     a 100  161  38   92    1.1  0.9  1.9  0.3  0.9   fortsættes
```

Remarks: Uses alpha numeric characters ('spår') in the numerical data section to signal traces.

Supplement B: NORFOODS Address List 1988

Country, representative	Work		
Denmark Anders Møller	The National Food Agency Møtkhøj Bygade 19 DK–2860 Søborg Denmark	Telephone Telex Telefax	+45 (0) 31 69 66 00 15 298 foodin dk +45 (0) 39 66 01 00
Finland Maarit Ahola	Food Research Program Ministry of Agriculture and Forestry Viikki 22 A SF–00710 Helsinki Finland	Telephone Telex	+358 (9) 0 372 088 122 352 hymk sf
Iceland Olafur Reykdal	Agricultural Research Institute Keldnaholt IS–112 Reykjavík Iceland	Telephone Telex	+354 (9) 1 82 230 2307 ISINFO-IS
Norway Arnhild Haga-Rimestad	Avd. for kostholdsforskning Postboks 1117, Blindern N–0317 Oslo 3 Norway	Telephone	+47 (0) 2 45 42 10
Sweden Lena Bergström	National Food Administration Nutrition Section Box 622 S–715 26 Uppsala Sweden	Telephone Telex Telefax	+46 (0) 18 17 57 30 76121 slvups s +46 (0) 18 10 58 48

Anders Møller, PhD, The National Food Agency, Mørkhøj Bygade 19,
DK–2860 Søborg (Denmark)

Simopoulos AP, Butrum RR (eds): International Food Data Bases and Information Exchange. World Rev Nutr Diet. Basel, Karger, 1992, vol 68, pp 121–135

Past and Present Activities in Food Composition Tables in Latin America and the Caribbean Islands

Ricardo Bressani, Marina Flores

Division of Food Science and Agriculture, INCAP, Guatemala City, Guatemala

Contents

Introduction

Before regional or national food composition tables existed, the Food and Agriculture Organization (FAO) of the United Nations published food composition tables in 1949 and 1954 for international use, based on analytical data from different parts of the world [1, 2]. These tables were intended to be used in countries that had no food composition data and to compare the nutrient content of food supplies in the different countries for the development of Food Balance Sheets. Recently, the important association between nutrient intake and disease as well as different consumption patterns among regions requires more detailed national food composition tables.

Any nutrition-related project requires the study of the dietary aspects of the subjects or population groups, where foods consumed are converted into values of nutrients from food composition tables. However, because of the great variability of the nutrient content and bioavailability in raw, processed, and cooked foods, it is critical to know whether the values are appropriate for the study. Food composition tables have been prepared by different authorities who each adopt different criteria for the methodologies and treatment of food samples according to the techniques practiced in their laboratories. Therefore, it is necessary and extremely important to study the introductory text and explanation notes given in these tables to document sources and determine the procedures followed in the compilation of the data.

The most extended use of food composition tables is the evaluation of the food patterns of people from different cultures or regions. National tables are likely to have the most reliable values for the foods in a given country. However, differences in the chemical composition of the same food among neighboring countries are often not important; especially if allowances are made for variation in water content. Consequently, food composition data compiled regionally may be satisfactory. Discussed below is the development of national Latin American and the Caribbean Islands food composition tables, their status, and future prospects.

Historical Perspectives Related to Food Composition Tables in Latin America

Before 1940, scientists working in university laboratories and the National Nutrition Institutes in Latin America began researching the chemical composition of foods. In Chile and Argentina, well-known nutritionists focused their studies on the nutritive value of regional foods and published their results in the first National Food Composition Tables. The first edition of the Chilean tables was published in 1961, with 112 food items [3]. The second edition was printed in 1965 and included 133 food items. After a series of updated tables, the sixth edition was published in 1979 and included 379 food items. Since then, complementary publications on various aspects of food composition have been published, including a 1945 publication from Argentina [4]. In 1954, the first Food Composition Table for Brazil was published, and emphasized the vitamin content of Brazilian foods [5]. Updates of this table were published in 1974 and 1982.

In 1965, analytical chemists from the Ecuadorian National Nutrition Institute in collaboration with biochemists from the Massachusetts Institute of Technology (MIT) prepared the first composition table of Ecuadorian foods [6]. Although some nutrient data on Mexican foods were published before 1940, the National Institute of Nutrition in Mexico collaborated with MIT to initiate a systematic analysis of the nutrient content of Mexican foods and published the table in 1951. The Colombian National Institute of Nutrition was founded in 1944, and one of its first focuses was the analytical study of foods. This work formed the basis for the Food Composition Tables for Colombia [7]. In Venezuela, the biochemical laboratories of the National Nutrition Institute dedicated their initial efforts to analyses of local foods, which led to the development of the first Venezuelan food composition table [8].

In Peru, seven documents were published from 1950 to 1975 on the nutrient content of the more commonly consumed Peruvian foods. Other documents include a table on fatty acid content, minerals, and indigenous foods consumed by rural inhabitants and minority groups [9]. Similar activities have been conducted in Bolivia [10].

Since 1952, nutrition researchers in Cuba have been very active in studying the edible flora and fauna of the country and compiling nutrient analyses of their foods. Local scientists in food science and nutrition in collaboration with food scientists from MIT presented their results in three different publications. In 1955, 115 samples of vegetable foods were analyzed; in 1957, results from analyses of 137 other vegetable foods were published, and in 1963 , 106 samples of foods collected in different parts of the island were analyzed [11–13].

English-speaking countries of the Caribbean Islands have developed food composition tables from data analyzed and compiled in the United States (US), FAO tables, and analytical values from the laboratories of the Jamaica Nutrition Institute [14].

Vegetable foods from the Central American area that includes Guatemala, Honduras, El Salvador, Nicaragua, and Costa Rica were the subject of an extensive and careful cooperation food analysis project between the laboratories of MIT and the United Fruit Company. Coordinated by MIT, the research team included a botanist who identified and classified plants and local chemists who analyzed plants from different localities of the Central American countries from 1944 to 1946. Their findings were presented in a series of publications in the *Journal of Food Research* of the USA in 1949 and 1950 [15]. Food sample collection methods from different farms and markets in the area and details concerning the analytical methods used in the MIT laboratories were described in the reports. To facilitate the identification of vegetable foods, information on plant part, size, color, and maturity was provided. Local cooking practices were also described.

The information, produced at MIT, was presented to the Instituto de Nutrición de Centro América y Panamá (INCAP) to provide a framework for analysis of foods and dietary studies in Central America. Since this presentation, food consumption studies have continued in Central American countries to assess the nutritive value of the inhabitants' dietary patterns. In the process, many other food items, including animal foods, were identified and analyzed by INCAP. In 1952, nutritionists and food scientists at INCAP began research on food habits of rural and urban populations in Central America and on the effects of indigenous food processing on nutrient composition of commonly consumed foods. With the analytical information available in 1953, the first Composition Table for Central America was published [16]. Most of the figures presented in the tables are averages of several samples of the same foods produced in different localities. Values that deviated greatly from the averages were not included. Carbohydrate values were calculated by difference, and specific factors were applied to obtain total calories for each food. The calorie and nitrogen conversion factors for protein were adopted from figures provided by the United States Department of Agriculture (USDA) [17], taking into account the digestibility of the foods [18].

During 1950–1960, INCAP conducted a dietary survey using 7-day dietary records. This information provided valuable data on edible portions of foods. Many of the national food composition tables, as well as regional tables, express nutrient values in terms of the edible portion per 100 g. The FAO international food composition tables present the values in terms of the edible portion and include the gross or as purchased weight of the foods. The edible portion of foods is obtained at the laboratories by applying correction factors based on the proportion of refuse and water content.

Results of food consumption studies carried out from 1950 to 1965 in Central and South America helped to identify foods needing analysis. At the end of this period, most of the Latin American countries published their own national food composition tables. Food analyses at the chemical laboratories of INCAP and others included special studies on processed and native foods such as corn tortillas, bread, different kinds of cheese, and many typical dishes. Analyses were also conducted on the basic staple foods such as maize, beans, and sorghum. With all the data from MIT and

INCAP, a new food composition table was prepared for the Central American countries in 1960 [19]. This table included the common and scientific name of each item, percentage of discarded part, water content, calories, and protein, fat, carbohydrates, fiber, ash, calcium, phosphorus, iron, vitamin A activity, thiamine, riboflavin, niacin, and ascorbic acid content in 100 g of edible portion. The acceptability of and great demand for this table was because of its attractiveness, clear presentation, the classification of nutrient data according to the food groups, and inclusion of nutritive values for local foods for each country.

State of Food Composition Tables

As discussed above, most of the available food composition tables were developed in the early 1940s. The Latin American food composition table, published in 1961, contains 716 food items [20]. The number of food items in the Latin American tables varies from 155 in El Salvador [21] to 1,648 in Argentina (table 1). Table 2 provides information on the nutrients

Table 1. Food composition tables in Latin America

Country	Date of first edition	Number of food items[1]
Central America	1960	314 (1971)
Latin America	1961	716 (1961)
Argentina	1935–1942	1,648 (1945)
Bolivia	1966	645 (1984)
Brazil	1954	–
Caribbean (English)	1974	799 (1986)
Colombia	1944	294 (1978)
Chile	1961	378 (1985)
Dominican Republic	1964	572 (1985)
Ecuador	1954	586 (1965)
Mexico	1940	391 (1983)
Peru	1960	460 (1975)
Uruguay	1949	53 (1949)
Venezuela	1950	484 (1983)
El Salvador	1989	155 (1989)
Cuba	–	238 (1985)

[1] In last edition.

Table 2. Information on food composition available in Latin American food composition tables (table from country or region)

Component	Latin America	Argentina	Bolivia	Caribbean	Central America	Chile	Colombia	Cuba	Dominican Republic	Ecuador	El Salvador	Mexico	Peru	Uruguay	Venezuela
Proximate composition	yes	yes*	yes	yes*	yes	yes	yes	yes*	yes	yes	yes	yes	yes	yes	yes
Energy	yes	yes	yes	yes	yes	yes	yes	yes	yes	yes	yes	yes	yes	yes	yes
Calcium	yes	yes	yes	yes	yes	yes	yes	yes	yes	yes	–	yes	yes	yes	yes
Iron	yes	yes	yes	yes	yes	yes	yes	yes	yes	yes	–	yes	yes	yes	yes
Phosphorus	yes	yes	yes	–	yes	yes	yes	yes	–	yes	yes*	–	yes	–	yes
Sodium	–	yes	yes	yes	–	yes	–	–	–	–	–	–	yes*	–	yes*
Potassium	–	yes	yes	yes	–	yes	–	–	–	–	–		yes*		yes
Zinc	–	–	–	yes	–	yes	–	–	–	–	–		yes		
Trace elements	–	yes	–	–	–	yes	–	–	–	–	–		yes		
Carotene	–	yes	–	–	–	–	–	–	–	yes	–		yes		
Vitamin A	yes	yes	yes	yes	yes	yes	yes	yes	yes	–	–	yes	no		yes
Vitamin B_1	yes	yes	yes	yes	yes	yes	yes	yes	yes	yes	–	yes	yes		yes
Vitamin B_2	yes	yes	yes	yes	yes	yes	yes	yes	yes	yes	–	yes	yes		yes
Niacin	yes	yes	yes	yes	yes	yes	yes	–	yes	yes	–	yes	yes		yes
Folic acid	–	–	–	yes	–	–	–	–	–	–	–		no		
Vitamin B_{12}	–	–	–	yes	–	–	–	–	–	–	–		no		
Vitamin B_6	–	–	–	–	–	–	–	–	–	–	–		no		
Vitamin C	yes	yes	–	yes	yes	yes	yes	yes	yes	yes	yes*	yes	yes		yes
Fatty acid	–	–	–	yes	–	yes	–	–	–	–	–		yes		
Amino acids	–	–	–	essential	–	yes	–	–	–	–	–	yes	some		
Cholesterol	–	yes	–	yes	–	–	–	–	–	–	–		no		
Cellulose	–	yes	–	–	–	–	–	–	–	–	–		no	yes	
Oxalic acid	–	some	–	–	–	–	–	–	–	–	–		no		
Purinas and uric acid	–	some	–	–	–	–	–	–	–	–	–		no		
Fluoride	–	–	–	–	–	yes	–	–	–	–	–		yes		
HCN	–	–	–	–	–	yes*	–	–	–	–	–				
Observation		*No ash		*No ash		*Beans		*No ash			*Fruits and vegetables		*Special table		*Special table

that have been analyzed for foods listed in the available food data tables [4, 6, 8, 10, 19, 22–41]. Some tables contain less common analytical information, including purines and uric acid, possibly because of the high intake of animal products in that region.

The types of food items vary from table to table, particularly those foods that are inherent to a particular country. Most tables follow the same food groupings: cereal and cereal products, vegetables, fruits, dry food legumes and their products, almonds and nuts, dried seeds, sugars and syrups, meats and fowl, eggs, fish and shellfish, milk and dairy products, fats and oils, beverages, and miscellaneous foods. Some include mixed dishes and special food items. Nutrient data include moisture content, protein, fat, crude fiber, ash, carbohydrate, calories, calcium, phosphorus, iron, thiamine, riboflavin, niacin, vitamin A, carotenes, and vitamin C. Only a few tables have information on trace elements or amino acid and fatty acid content. A recent table from Chile, however, has very complete data on fatty acids for animal, fish, and vegetable oils [24]. Various countries have published a table on the content of sodium and potassium in foods [25]. Some of the analytical methodologies used to produce these data are old, but efforts are being made to update the methodology in a number of countries.

It should be noted that most information comes mainly from nutrition institutes. Data exist from a number of university laboratories and/or quality-control laboratories, but efforts have not been made to incorporate those data into the existing data base. For example, although the carotenoid composition of Brazilian foods is available in a number of publications [26], it cannot be found in the food composition tables. This area requires more attention if Latin American data bases are to be updated and improved.

Updating and Other Needs

Over the past 25 years in Latin America, significant changes in analytical methodologies and agricultural technology have taken place [27], and food composition tables must be updated to represent these advances. Examples of these technological advances include the production of hybrids for maize and the intensive use of compounded and medicated feeds for animal products. Agricultural practices now include greater use of chemical fertilizers to control weeds and plant diseases. Similarly, there has been a steady increase in processed food products that use new techniques that may affect the nutrient content of foods. Lime-treated maize tortillas or arepas were mainly a home process some 25–30 years ago, but today tortillas are made from processed maize flour and sold packaged in the supermarket. In addition, beans are being used in other products such as cooked bean flours, canned whole beans, or canned bean paste. The same tendency has taken place with animal food products.

In Latin America, the migration to urban centers has led to the adaptation of new lifestyles and dietary habits. People have a greater choice of ready-to-eat foods or fast-foods for which no analytical values are available. In urban centers, an increasing number of street food vendors sell many types of foods, most of which have not been analyzed. Although processed foods are packaged in different types of materials, information related to changes in food composition upon storage is not available. In summary, many factors must be considered as efforts to update food tables are initiated.

First, a system designed to select and incorporate new analytical data into an existing data base needs to be developed. Standards for coding and describing food data should be included in light of the large number of different food items found in the various regions of Latin America. This system should include standardized sample collection methods, supporting data for sample identification, descriptions of processes – mainly those of autochthonous origin, improved and accepted analytical techniques, reference standards, and reporting method.

Updating the quality of present food composition tables also depends on the needs of the users. Although vitamin A deficiency is quite common among population groups in Latin America, the information available on carotenoid content in vegetable products is poor [26]. The deficiency problem should be solved by a greater consumption of food sources rich in carotenes; however, because of the lack of data, the problem is being solved through fortification programs. Iron deficiency is another problem, yet little attention has been given to the validity of the iron values in the tables. With present changes in dietary habits, including diets of low fiber content, attention should be given to dietary fiber content in the main foodstuffs consumed by the population. Although many people base their diet on root and tubers, carbohydrate content is still being obtained by difference as opposed to analysis. Furthermore, data has been produced in various laboratories, but no efforts have been made to retrieve them, select them, and incorporate them in data bases. Efforts to update and improve these tables must be organized and approached systematically. One organization that has played an important role in this area is LATINFOODS.

LATINFOODS

In November 1986, the First Conference on Food Composition was held at INCAP, sponsored by the International Development Research Centre (IDRC), the United Nations University (UNU), and the US Agency for International Development (AID). The main objectives of the meeting were: (1) to review the state of knowledge of food composition tables for the individual countries and for the region; (2) to propose programs aimed at increasing the usefulness of and upgrading present tables in terms of quantity and quality of analytical data, and (3) to develop a network of people and institutions interested in food composition tables through the development of LATINFOODS [28].

At this conference, several reports were presented on the historical perspectives and compilation of available food composition data and the country's need for new data [7, 14, 29–41]. These reports strongly indicated that data are available and should be obtained, selected, and incorporated into data bases. To reach the second objective, conclusions and recommendations were requested from three groups: users of food composition data; compilers of food composition values, and data producers. The users working group indicated that present tables were incomplete because they did not contain information on many indigenous foods or on new foods from the food industry. They recommended that international programs donate food composition data. Because most values are given for raw foods, the need for nutrient values for foods as they are commonly consumed was emphasized. With respect to nutrients, iron and vitamin A values received some priority, indicating that present values are inconsistent and incomplete. Other nutrients include sodium and potassium values as well as zinc and iodine, fatty acids, dietary fiber and specific carbohydrates. Many participants expressed interest in citing values for polyphenolic compounds, oxalates, and phytates. The data producers working group reinforced the recommendations of the users group, which included the need for updated equipment and training and guidelines for sampling and selection of data. In addition, the need for collaborative studies and increased communication and interchange of information was discussed.

The data compilers agreed on the need for development of guidelines for data selection and appropriate reporting as well as strong collaboration with those producing and using food composition data.

There was agreement among the three groups to create and implement LATINFOODS. The first step in this process was to recover available data and develop a set of criteria for data selection. The criteria should constitute a shared data base that could serve as a basis for developing specific tables for local needs. Close collaboration and communication between data users and producers would be necessary to accomplish these steps.

National groups were created, consisting of representatives from various institutions and disciplines. The center of the network would be located at INCAP, with a committee consisting of a coordinator, four subregional representatives from Mexico, Venezuela, Brazil, and Chile, and the president of the Latin American Society of Nutrition.

Present objectives of LATINFOODS include identifying sources of food composition data, developing quality criteria for selection of data; promoting data generation; acquiring and disseminating new analytical data; facilitating the access, production, and interchange of data; and developing activities that will keep the concept alive.

Achievements and Prospects

Although funding issues have not been resolved, LATINFOODS has had numerous achievements since 1986. Eight national groups have been established in Argentina, Brazil, Chile, Bolivia, Ecuador, Venezuela, Costa Rica, and Nicaragua. Despite economic limitations, these groups hold regular meetings and submit proposals to the Coordinator General that are related to initiating specific regional activities.

With the help of the International Program in the Chemical Science (IPICS) and the Chemistry Center of the University of Lund, Sweden, a short course on dietary fiber analysis was conducted at INCAP in February 1988 and attended by 16 participants from Mexico, Central America, Colombia, and Ecuador. From this analytical course, arrangements have been made for a collaborative study supported by IPICS on dietary fiber analysis. Its success will contribute significantly to the future development of LATINFOODS.

A small grant administered by the LATINFOODS headquarters was provided to three national groups in Central America to produce new data on foods selected by the national groups. From this small effort, a table of food data from El Salvador has been published [21], and one from Costa Rica will soon become available. In addition, a form for compiling data accompanied by an instruction book for collection of information was developed at INCAP. These forms provide space for a detailed description of individual food and raw or processed food composites.

A second LATINFOODS meeting was held in November 1988 in Chile where various speakers discussed current analytical methodology in Latin America. Present capabilities include fatty acid analysis, mineral analysis, and dietary fiber analysis. However, analytical methods for vitamins, carotene, and carbohydrate still need improvement. Strategies for obtaining necessary funding for and conducting collaborative studies throughout Latin America were also discussed.

The concept of LATINFOODS was presented at the VIth Latin American Meeting in Food Science and Technology, October 1988, in Bogota, Colombia, and at the VIIth Latin American Nutrition Congress, November 1988, in Viña del Mar, Chile. These presentations emphasized the importance and significance of LATINFOODS and increased the number of contributions to the concept. In addition to a LATINFOODS newsletter, a section on food composition in the journal *Archivos Latinoamericanos de Nutrición,* has been established. These provide new information related to dietary fiber content of cereal grains, food legumes, and vegetables, fatty acid content of various oils, trace mineral content of processed basic staple foods, and carotene content of local fruits such as pejibaye and native vegetable crops.

Early in 1990, a third meeting of LATINFOODS was held in San Jose, Costa Rica. Highlights of the conference included uses of food composition data by the food industry, needs for food labeling data, the effects of processing on composition, and the use of computers in the development of data bases.

Latin American food composition tables were originally developed to interpret nutrition surveys. Although nutritional deficiency problems still prevail in many Latin American countries, problems because of nutrition-related diseases such as cancer and cardiovascular disease are increasing. Food composition tables must be updated to reflect these new concerns. Food and nutrition surveillance activities can be a useful approach for the collection and analysis of new data to enrich present tables. Many nutrition intervention studies need updated nutrient intake data to measure the variables. New multidisciplinary research teams throughout Latin America are becoming active again because of technological advances, including new equipment, and the use of computer programs. Other activities that are fostering interest in developing food data bases include new nutritional evaluation concepts such as chemical scores for protein that are based on essential amino acid content and digestibility, the presence of foods or ingredients in canned foods for which no analytical values are available such as the oil in canned fish, food products derived from new processing technologies, food labeling, and food fortification. LATINFOODS plays an important role in the renewed interest in food composition data.

Conclusion

This communication reviews the past, present, and future of food composition tables in Latin America and the Caribbean Islands. Food composition tables were originally constructed from analytical values obtained during 1940–1960. The participation of food and nutrition scientists from MIT was essential for the development of tables from Mexico, Cuba, Central America, and Ecuador. In addition, national scientists were responsible for tables developed in Venezuela, Chile, and Argentina.

The identification and collection of data from food items not available in existing food composition tables was first made possible by nutrition surveys that defined dietary patterns, nutrient intake, and nutrient deficiencies. The first attempt to integrate available data from nutrition institutes can be traced to the development of food composition tables for use in Latin America. The nutrients listed include energy, proximate composition, and limited number of macro- and micronutrients. Some tables contain data on sodium, potassium, fatty acids, and amino acids. Not all data are original to each country table; many values are obtained from other tables, including those from the US or FAO.

Because of changes in analytical techniques, agricultural and processing technology, dietary habits, and the lack of standardization among food composition tables, LATINFOODS was created in 1986 to maintain and update food composition data in Latin America and the Caribbean Islands. In the future, LATINFOODS will continue to have a major role in creating more representative and useful food composition tables.

References

1 Chatfield C: Food and Agriculture Organization of the United Nations. Washington, USA, 1949. (Calories, protein and fat.) FAO Nutritional Studies No 3.

2 Chatfield C: Food and Agriculture Organization of the United Nations. Rome, Italy, 1954. (Food composition tables – minerals and vitamins – for international use.) FAO Nutritional Studies No 11.

3 Tabla de Composición Química de Alimentos Chilenos. Cátedra de Bromatología, Facultad de Química y Farmacia, Universidad de Chile, 1961.

4 Tablas de la Composición Química de los Alimentos. Materias Primas y Preparaciones Alimenticias. 4a ed. Instituto Nacional de la Nutrición, Buenos Aires, Argentina, 1945.

5 Cramer ER, et al: Valor vitamínico de alimentos Brasileiros (Colecao Estudos e Pesquisas Alimentar). Rio de Janeiro, SAPS, 1954.

6 Tabla de Composición de los Alimentos Ecuatorianos. Ministerio de Previsión Social y Sanidad, Instituto Nacional de Nutrición, Quito, Ecuador, 1965.

7 Fajardo L, Gracia de RB, Lareo L: Informe de Colombia. Arch Latinoam Nutr 1987; 37:751–768.

8 Tabla de Composición de Alimentos para Uso Practico. Revisión 1983. Ministerio de Sanidad y Asistencia Social, Instituto Nacional de Nutrición, Dirección Técnica, División de Investigaciones en Alimentos, Publ No 42, Caracas, Venezuela, 1983.

9 Alarcón NH, Hernández FE, Faching de Forton A, Robles GN: La Composición de los Alimentos Autóctonos. Perú, 1977, 1a ed. Ministerio de Salud. Institutos Nacionales de Salud. Instituto de Nutrición. Laboratorio de Bromatología y Bioquímica.

10 De Ibañez JT, Inofuentes DG: Tabla de Composición de Alimentos Bolivianos. Ministerio de Previsión Social y Salud Pública. División Nacional de Nutrición. Laboratorio Bioquímica Nutricional. La Paz, Bolivia, 3a ed, 1984.

11 Navia JM, López H, Cimadevilla M, Fernández E, Valiente A, Clement ID, Harris RS: Nutrient composition of Cuban foods. I. Foods of vegetable origin. Food Res 1955;20:97.

12 Navia JM, López H, Cimadevilla M, Fernández E, Valiente A, Clement ID, Harris RS: Nutrient composition of Cuban foods. II. Foods of vegetable origin. Food Res 1957;22:131.

13 López H, Navia JM, Clement D, Harris RS: Nutrient composition of Cuban foods. III. Foods of vegetable origin. J Food Sci 1963;28:600–610.

14 Patterson AW: English-speaking Caribbean region report. Arch Latinoam Nutr 1987;37:772.

15 Munsell H: Composition of food plants of Central America. Journal of Food Research of the USA: I. Honduras Food Research. II. Guatemala Food Research 1949;14:144. III. Guatemala Food Research 1950;15:16. IV. El Salvador Food Research 1950;15:263. V. Nicaragua Food Research 1950;15:355. VI. Costa Rica Rood Research 1950;15:379. VII. Honduras Food Research 1950;15:421. VIII. Guatemala Food Research 1950;15:439.

16 Instituto de Nutrición de Centro América y Panamá. Tercera edición de la Tabla de Composición de Alimentos de Centro América y Panamá. Boletín de la Oficina Sanitaria Panamericana. Supl, No 1, 1953.

17 Watt K, Merrill AL: Composition of foods raw, processed, prepared. US Department of Agriculture, Agriculture Handbook 8, Washington, D.C. pp 1–147, 1950.
18 Jones DB: Factors for converting percentages of nitrogen in food and feeds into percentages of protein. USDA. Cir 183, p 22, 1941.
19 Flores M, Flores Z, García B, Gularte Y: Tabla de composición de Alimentos de Centro América y Panamá, 4a ed. Guatemala, C.A. Instituto de Nutrición de Centro América y Panamá. Guatemala, enero 1960.
20 Wu Leung W-T, Flores M: Tabla de Composición de Alimentos para Uso en América Latina. INCAP-ICNND, Junio 1961. INCAP, Guatemala.
21 Merino JG: Composición Química de Alimentos Populares en El Salvador. UCA Editores. San Salvador, El Salvador, C.A., 1a ed. 1989.
22 Bressani, R: The data required for a food data system Food Nutr Bull 1989;5:69.
23 Collazos CHC, White PL, White HS, Viñas TE, Alvistur JE, Urquieta AR, Vásquez GJ, Dias TC, Quiroz MA, Roco NA, Hegsted DM, Herrera AN, Fachin RA, Robles GN, Hernández FE, Bradfield RB: La Composición de los Alimentos Peruanos. Ministerio de Salud, Institutos Nacionales de Salud. Instituto de Nutrición, Lima, Perú, 5a ed, 1975.
24 Masson SL, Mella RMA: Materias Grasas de Consumo Habitual y Potencial en Chile. Composición en Acidos Grasos, 1a ed. Lab de Química y Bioquímica de Alimentos. Facultad de Ciencias Básicas y Farmacéuticas. Universidad de Chile, 1984.
25 Tabla de Contenido Promedio de Na y K en Algunos Alimentos Comunes. Ministerio de Sanidad y Asistencia Social. Instituto Nacional de Nutrición. Dirección Técnica, División de Investigaciones en Alimentos. Caracas, Venezuela, April 1982.
26 Rodríguez Amaya DB: Critical review of provitamin A determination in plant foods. J Micronutr Anal 1989;5:191–225.
27 National Research Council. Lost crops of the Incas. Little-known plants of the Andes with promise for worldwide cultivation. National Academy Press, Washington, D.C.
28 Memorias de la Primera Reunión Sobre Tablas de Composición de Alimentos LATINFOODS. Arch Latinoam Nutr 1987;37:607.
29 Tabla de Composición de Alimentos. Cuba. Instituto de Investigaciones para la Industria Alimenticia (MINAL). Instituto Nacional de Higiene de los Alimentos (MINSAP). Habana, Cuba, 1985.
30 Mussgay, B: Tabla de composición de Alimentos para Uso en República Dominicana. Servicio Alemán de Cooperación Social – Técnica (DED). Instituto para el Desarrollo del Suroeste (Indesur). Azua RD. Enero 1985. República Dominicana.
31 Tejerina de Ibañez J, Linares GS, Feraudi M: Tabla de Composición de Alimentos Bolivianos. Arch Latinoam Nutr 1987;37:714.
32 Hernández M, Chávez A, Bourges H: Valor Nutritivo de los Alimentos Mexicanos. Tablas de Uso Práctico. Instituto Nacional de la Nutrición (INN). División de Nutrición L-12, 9a ed. Mexico, 1983.
33 Composición Química de Algunos Alimentos Consumidos en el Uruguay. Comisión de Alimentación. República Oriental de Uruguay. Montevideo, Uruguay, 1949.
34 Lajolo F, Vanucchi H: Tabelas de Composiçao de Nutrientes em Alimentos é Situaçao no Brasil e Necessidades. Arch Latinoam Nutr 1987;37:702.

35 Masson LS, Araya H, Mella MA: Estado Actual de las Tablas de Composición de Alimentos en Chile. Arch Latinoam Nutr 1987;37:683.
36 Closa SJ, de Portela MLPM, Sambucetti ME, Longos E, Schor I, Carmuega E: Informe sobre estado actual, interés y limitaciones existentes con referencia a Tablas de Composición de Alimentos en la República Argentina. Arch Latinoam Nutr 1987;37:694.
37 Jaffé WG, Adam G: Utilización de la Tabla Oficial de Composición de Alimentos en la Actualidad. Informe de Venezuela. Arch Latinoam Nutr 1987;37:730.
38 Rondón H, Jiménez de Sánchez R: Uso de la Tabla de Composición de Alimentos en la República Domincana. Arch Latinoam Nutr 1987;37:769.
39 Alvarado J de D, Gallegos Espinoza S: Datos de Composición de Alimentos en El Ecuador. Arch Latinoam Nutr 1987;37:723.
40 Bourges H, Valencia M: Análisis de la Composición de los Alimentos en México. Antecedentes, Situación Acutal y Perspectivas. Arch Latinoam Nutr 1987;37:785.
41 Hernández FEM: Informe de Perú. Composición Química y Valor Nutritivo de Alimentos Nativos Andinos. Arch Latinoam Nutr 1987;37:719.

Ricardo Bressani, PhD, Institute of Nutrition of Central America and Panama, Carretera Roosevelt, Guatemala City 01011 (Guatemala)

Simopoulos AP, Butrum RR (eds): International Food Data Bases and Information Exchange. World Rev Nutr Diet. Basel, Karger, 1992, vol 68, pp 136–156

Food Composition Tables and Food Composition Analysis in East Europe

Jerzy H. Dobrzycki[a], *Maria Los-Kuczera*[b]

[a] Agrotechnology and Veterinary Research Center of Polish Academy of Sciences, and [b] National Institute of Food and Nutrition, Warsaw, Poland

Contents

Introduction

Over the past 10 years, computerized food composition tables have been developed and compiled in Europe and other regions of the world. A complete list of these data bases was compiled by the International Network of Food Data Systems (INFOODS) [1] and funded by the US National Cancer Institute under the auspices of the United Nations University. Data bases at the regional level include LATINFOODS, ASIAFOODS, OCEANIAFOODS, and NORFOODS. More than 20 countries in Europe, both West and East, have been involved in the work of EUROFOODS [2]. Although the meetings (Wageningen 1983, Norwich 1985, Warsaw 1987, and Uppsala 1989) proved to be very useful, much work remains to be done in this area, mainly because of the diversified nutrient values and food descriptions in these international food composition tables [3, 4]. The information presented below describes the published East

European Food Composition (EEFC) tables that are currently in use. The possibility of using these tables in an internationl computerized information system also is discussed.

Food Composition Tables in East Europe

All of the East European countries except Albania have detailed food composition tables. Table 1 lists these tables including a 1965 and 1988 publication for Czechoslovakia (CS) II, 1988 publications for Hungary (H) and Poland (PL) and older publications of tables from Bulgaria (BG), the German Democratic Republic (GDR), and Yugoslavia (YU). The number of food products range from 500 to 1,000 for YU, CS, GDR, and PL and from 1,600 to 1,700 for the Soviet Union (SU) and H. Table 1 also lists the author, edition, publisher, and number of pages for these tables.

The main difficulty other foreign countries have in using these tables is that most of the food composition tables have been written in the national languages. Only the 1988 tables from CS have an English introduction and a Czech-English, English-Czech index of products thus necessitating the inclusion of systematic (Latin) names for plants, fish, etc., to assist the user with other regional tables.

Although most food composition tables categorize their food products into standard groups, there are significant differences in the content of these groups among East European tables. For example, in the CS, BG, and SU tables, there are only 8–12 product groups while in the H, GDR, and PL tables, 16, 25, and 39 groups are shown, containing from 1 to 287 products. The products' assignment to groups is not uniform. For example, eggs are placed in 'poultry and poultry products' (BG, SU), 'milk and milk products' (CS), and as their own group (GDR, H, PL). The same is true of mushrooms and potatoes – they are placed either in a separate group or included in the vegetable group.

The differences in the classifications of food groups hinder comparisons among tables. In figure 1, for comparison purposes, we have made a division into nine main groups – meat, milk, fat, cereal, vegetables, fruit, sugar, prepared foods, and miscellaneous products. The proportion of total products contained in the tables has been calculated for each group. More specifically, the groups most frequently presented in the EEFC tables (BG, GDR, H, PL) are 'meat and meat products', which account for about 25% of all food items, and 'prepared foods', accounting for 20–40%. The vast

Table 1. Tables of food composition published in East European countries

Author	Title		Year of publication/ current edition	Publisher (country)	Number of pages	Number of food products
	in national language	in English				
Tashev and Shishkov [5]	Tablici za sostava na bolgarskite chranitielni produkti	Tables of the Composition of Bulgarian Food Products	1975/2nd	Medicina i fizkultura publ. House, Sofia (BG)	349	1,113
Krondlova-Skopkova and Smrha [6]	Tabul'ky vyzivnych hodnot potravin	Tables of Nutritive Value of Foods	1965/3rd	Statni Zdravotnicke Nakladatel'stvi Praha (CS)	576	642
Kajaba and Smrha [7]	Tabul'ky zlozenia a vyzivovych hodnot pozivatin	Tables of the Composition of Food Products	1985/2nd	Slovenske Pedagogicke Nakladatel'stvo, Bratislava (CS)	102	624
Strmiska [8]	Pozivatinove tabul'ky. I. Potravinove suroviny	Food Tables. I. Primary Foods	1988/1st	Vyskumny Ustav Potravinarsky, Bratislava (CS)	189	646
Haenel [9]	Energie- und Nährstoffgehalt von Lebensmitteln. Lebensmitteltabelle	Energy and Nutrient Content of Foods. Food Composition Tables	1979/1st	VEB Verlag Volk und Gesundheit, Berlin (GDR)	896	841
Mohr and Pankrahts [10]	Kleine Lebensmitteltabelle	Short Food Composition Table	1985	VEB Verlag Volk und Gesundheit, Berlin (GDR)		280
Biró and Lindner [11]	Tapanyagtablazat /Tapanyagszukseglet es tapanyagosszetetel/	Food Composition Tables	1988/11th	Medicina Konyvkiado, Budapeszt (H)	261	1,740

Piekarska and Los-Kuczera [12]	Sklad i wartosc odzywcza produktow spozywczych Cz.I.	Composition and Nutritive Value of Food Products. Part I	1983/4th	Panstwowy Zaklad Wydawnictw Lekarskich, Warszawa (PL)	212	991
Los-Kuczera and Piekarska [13]	Sklad i wartosc odzywcza produktow spozywczych Cz. II–VII	Composition and Nutritive Value of Food Products. Parts II–VII	1988/1st	Panstwowy Zaklad Wydawnictw Lekarskich, Warszawa (PL)	496	ca. 650
Piekarska et al. [14]	Popularne tabele wartosci odzywczych zywnosci	Popular Tables of Nutritive Values of Food	1983/2nd	Panstwowy Zaklad Wydawnictw Lekarskich, Warszawa (PL)	59	512
Segal et al. [15]	Valoarea Nutritiva a produselor agroalimentare	The Nutritive Value of Agricultural Food Products	1983	Ed. Ceres, Bucuresti (Romania)		
Skurihin and Volgarev [16]	Chimiceskij sostav piscevych produktov. Kniga 1: Spravocnyje tablicy soderzanija osnovnych piscevych vescestv i energeticeskoj cennosti piscevych produktov	Chemical Composition of Food Products. Book 1: Composition Tables of Basic Nutrients and Energy Value of Food Products	1987/2nd	VO 'Agroprom izdat', Moskva (SU)	224	1,592

(Table 1 continued next page.)

Table 1 (continued)

Author	Title		Year of publication/ current edition	Publisher (country)	Number of pages	Number of food products
	in national language	in English				
Skurihin et al. [17]	As above, Kniga 2: Spravocnyje tablicy soderzanija aminokislot, zirnych kislot, vitaminov, makro- i mikroelementov, organiceskich kislot i uglevodov	As above, Book 2: Composition Tables of Amino acids, Vitamins, Macro- and Microelements, Organic Acids and Carbohydrates	1987/2nd	VO 'Agroprom izdat', Moskva (SU)	360	ca. 700
Skurihin and Sodernikov [18]	As above, Kniga 3: Spravocnyje tablicy soderzanija osnovnych piscevych vescestv i energoticeskoj cennosti bliud i kulinarnych izdelij	As above, Book 3: Composition Tables of Basic Nutrients and Energy Value of Ready-to-eat Meals and Culinary Products	1984/1st	Legkaja i piscevaja promyslennost, Moskva (SU)	328	516
Brodarec [19]	Tablice o sastavu i prehrambenoj vrijednosti namirnica i pica	Tables of Composition and Nutritive Values of Food and Beverages	1976/4th	Zavod za zastitu zdravlja SR Hrvatske, Zagreb (YU)	134	492

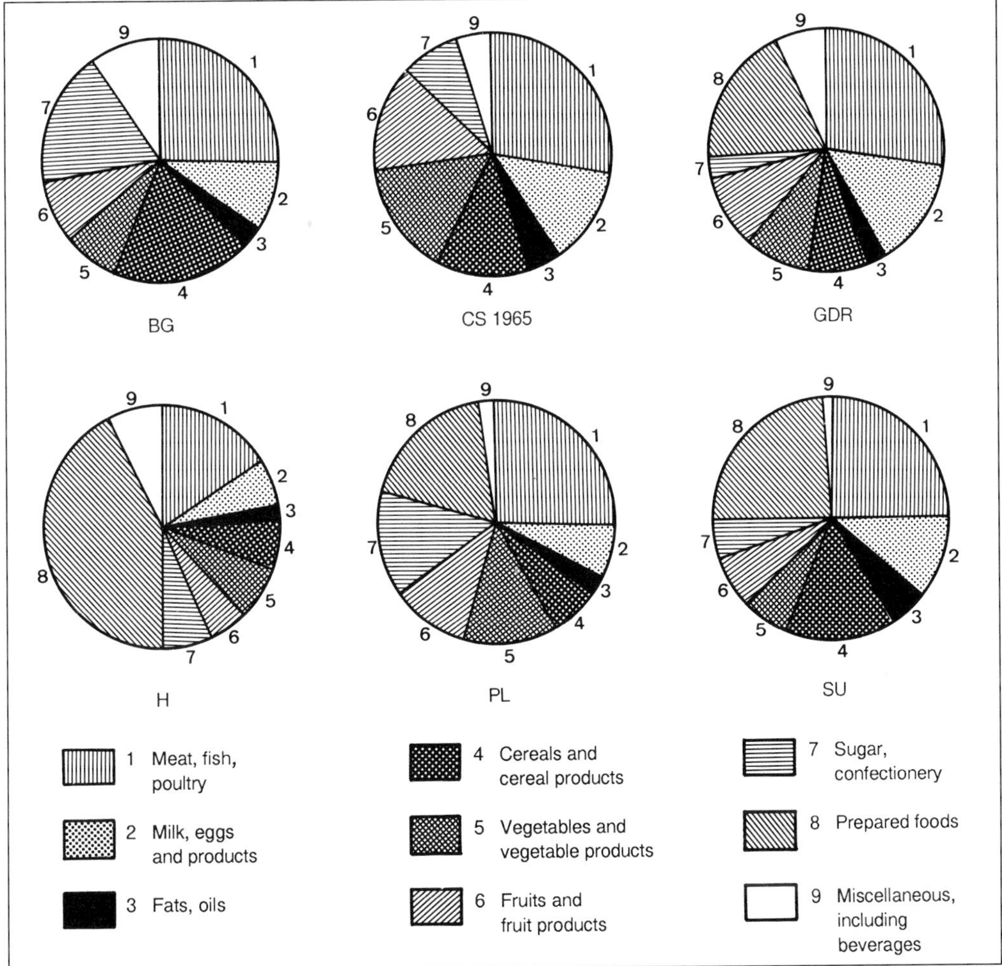

Fig. 1. Distribution of main food groups in EEFC tables.

differences found in the number of products in each group for each country may be a result of the following: (1) level of agriculture and food industry; (2) nutritional customs; (3) level of social education; (4) national research possibilities and availability of useful data.

The varying number of food products within the group is not the only obstacle to making a clear comparison. A more serious problem arises when comparing the number of nutrients, which can vary from 8 to 100 for individual countries. In PL and SU, two different books have been developed: one general book (PL I, SU I) and another (PL II, SU II) that gives detailed information on the same products. In BG, CS I, and H, detailed nutrient contents have been included for only some of the products. The comparison of such general and detailed books is shown in table 2. Food composition tables from different countries are briefly discussed below.

Bulgaria (BG)

The first Bulgarian food composition tables were published in the 1950s. Earlier editions included information on the composition of approximately 350 food items while the last edition, published in 1975, contains more than 1,100 food products. The foods listed in the tables are sequentially numbered. Fish, mushrooms, vegetables, fruit, and cereals also have the systematic (Latin) names. All data given in these tables refer to 100 g of the edible part (e.p.) of the food item (part I) and to 100 g of food as purchased (a.p.) (part II). Additionally, part III contains the amount of nutrient per 10, 20, 30 ... 100 g a.p. of the 285 most commonly supplied Bulgarian products. Nutrients given in the tables are listed in table 2 (column BG).

The Bulgarian food composition table is based on Bulgarian laboratories' analyses (indicated as boldface) as well as foreign literature (normal face). A '0' (zero) in the content column signifies that these constituents are not present in the food product. A zero in parentheses ('(0)') indicates that there is a very low content of a constituent, a dash ('–') means data are not available but does not exclude the presence of the nutrient in the foods. The tables include the amount of waste in grams and an index of products.

The following values are given in the appendix of the tables: (1) 18 amino acids for 93 products; (2) total saturated and unsaturated fatty acids for 67 products; (3) cholesterol content for 80 products; (4) 14 microelements and heavy metals for 107 products; (5) sugar, starch, and organic acids for 69 products. Preparation of an updated edition of the tables is under way.

Czechoslovakia (CS)

The first edition of the CS food composition tables was published in 1952 and updated by Krondlova-Skopkova and Smrha [6] in 1965. All data given in these food tables refer to 100 g e.p. and 100 g a.p. of food products. The 15 nutrients given in the tables are mentioned in table 2

Table 2. Nutritional factors reported in EEFC tables

Nutrient	Country										
	BG	CS I	CS II	GDR	H	PL I	PL II	SU I	SU II	SU III	YU
Energy, kcal	●	●	●	●	●	●		●		●	
kJ			●	●	●	●					
Water	●	●	●	●	●	●		●	●	●	●
Dry matter			●								
Protein, total	●	●	●[1]	●[1]	●	●		●	●	●	●
Amino acids							●		●		
Fat, total	●	●	●[2]	●	●	●		●		●	●
Fatty acids				●[2]			●		●		
Cholesterol				●			●				
Carbohydrates, total	●	●	●[3]	●	●	●	●	●	●	●	●
Oligosaccharides				●[3]			●				
Cellulose	●										
Fiber		●	●	●	●	●					
Organic acids								●		●	
Ash	●	●	●		●	●		●		●	
Minerals					[1]						
Na sodium	●		●	●	●		●	●	●	●	●
K potassium	●		●	●	●		●	●	●	●	●
Ca calcium	●	●	●	●	●	●		●	●	●	●
Mg magnesium	●		●	●	●	●		●	●	●	●
Fe iron	●	●	●	●	●	●		●	●	●	●
P phosphorus	●	●	●	●	●	●		●	●	●	●
Cl_1 chlorine			●	●			●		●		
I iodine			●	●			●		●		
NaCl sodium chloride			●								
Zn zinc			●	●	●		●	●	●		
Mn manganese				●	●		●		●		
Cu copper				●	●		●		●		
Co cobalt					●		●		●		
Cr chromium					●		●		●		
Ni nickel					●				●		
S sulphur							●		●		
F fluorine							●		●		
Mo molybdenum							●		●		
Se selenium							●		●[1]		

(Table 2 continued next page.)

Table 2 (continued)

Nutrient	Country										
	BG	CS I	CS II	GDR	H	PL I	PL II	SU I	SU II	SU III	YU
Vitamins					[1]						
A	•	•	•	•	•	•			•		•
Carotene	•			•	•	•		•	•	•	
D	•		•	•	•		•	•	•		
E	•		•	•	•		•		•		
K	•										
Thiamin/B_1	•	•	•	•	•	•			•	•	•
Riboflavin/B_2	•	•	•	•	•	•		•	•	•	•
PP/niacin	•	•	•	•	•	•		•	•	•	•
B_6			•	•	•		•	•	•		
Folic acid			•	•	•		•		•		
B_{12}			•	•	•		•		•		
C	•	•	•	•	•	•			•	•	•
Pantothenic acid			•	•	•		•	•	•		
Biotin			•		•		•		•		
Nutrients, total	22	15	35	36	8/33[1]	19	69	19	100	19	14

CS II: 1-total, plant and animal origin, 2-total and linoleic acid, 3-total and sucrose.
GDR: 1-total and purine content, 2-saturated and unsaturated, 3-mono-, di- and polysaccharide.
H: 1-for some products only.
SU II: additionally for some products: Ag, Al, B, Pb, Sr, Ti, V, Zn, Zr.

(column CS I). Part of the main tables (pages 163–433 in the original book) contains the nutrient composition of the 80 most popular food items expressed in grams. The following information is found in the tables: (1) 18 amino acids for 161 products; (2) saturated (g and %) and unsaturated (g and %) fatty acids for 89 products; (3) 6 minerals for 172 products; (4) 7 vitamins for 95 products; (5) energy, protein, fat, vitamin A, and vitamin C for 899 ready-to-eat, hot and cold dishes (soups, main dishes, salads, sandwiches, etc.).

The tables recently developed by the team of F. Strmiska from the Food Research Institute in Bratislava in 1988 are considered the most practical tables produced from the Central Food Data Bank. These are the only EEFC tables prefaced and indexed in English.

German Democratic Republic (GDR)

The current edition of these tables was edited by Prof. Haenel in 1979. Foods listed in the tables are numbered in a decimal classification system (group, subgroup, sub-subgroup, product). The values of each food item presented in these tables are separated into two columns: (1) first column – all entries are based on 100 g e.p.; (2) second column – refers to 100 g food a.p.

The average waste is expressed as a percentage. Most of the data in these tables are compiled from literature. Also included are intake levels of energy and nutrients recommended to the GDR population developed by the Nutrition Institute of the GDR Academy of Sciences and an alphabetical appendix of all foods in these tables.

Hungary (H)

The first Hungarian food composition tables, published in 1951, covered only 9 pages; the 11th edition, published in 1988, has 261 pages. All data are developed by the staff of Hungarian institutions; 70% of the data are from the National Institute of Food Hygiene and Nutrition.

Food groups and products are not numbered or indexed. The main table (number 40 on pages 102–162 of the original book) lists more than 1,700 foods and mixed dishes. The values for energy (kJ and kcal), protein, fat, carbohydrates, water, ash, and fiber are also included. Vitamins, minerals, and other nutrients are grouped as follows: (1) 13 vitamins for 538 products (table 43 in original publication) and 6 vitamins for another 287 products (table 42); (2) 12 minerals and alkalinity of ash for 449 products (table 44); (3) 18 amino acids for 45 products (table 47); (4) 14 fatty acids for 52 products (table 48); (5) cholesterol content for 41 products (table 49).

Poland (PL)

Polish food composition tables are prepared at the National Food and Nutrition Institute. Work in this area began in 1936 and 1938, with the vitamin tables [20]. Values in these tables reflect the work of many Polish and United States researchers, and they were based on scientific research related to vitamins [21, 22]. The first Polish food composition tables were published by Rudowska-Koprowska in 1954 [23], and by 1979 there were seven editions of the tables. In 1983, part I of the new tables was completed, and parts II–VII were created and updated in 1988. All values for food items are divided into two groups in the tables: the first group represents values calculated for 100 g e.p. of foods, and the second gives information for 100 g of products a.p. The percentage of waste also is given. Systematic (Latin) names of fish, mushrooms, nuts, fruit, vegetables, and yeasts are listed in the appendix.

An extensive introduction precedes each section of the food composition tables. There is information on the tables' arrangement, groupings of foods, and methods of nutrient analysis. Where possible, nutrient losses during preparation and processing of raw products are given. Parts I through VII are described as follows: (I) energy values, proximate constituents, 4 minerals and 5 vitamins for 991 food items; (II) 13 minerals for 650 food items; (III) 8 vitamins for 601 products; (IV) fatty acids, approximately 20 saturated, monounsaturated, and polyunsaturated acids, for 183 foods; (V) cholesterol for 183 products; (VI) amino acid composition and limiting amino acid factor of protein for 115 products in mg/g nitrogen, and 18 amino acids for 333 products; (VII) carbohydrates: monosaccharides, disaccharides, and polysaccharides in 123 products.

Priority has been given to Polish values rather than foreign data; however, in cases where it has not been possible to procure Polish data, values from foreign tables have been used. All data have source references at the end of each section.

A new edition of the food table was recently published [30]. The book contains data on energy and 78 nutrients (no empty spaces) for 224 food products and is intended to computerize programs that are used in epidemiological research.

Soviet Union (SU)

The Russian food composition tables were prepared by a special interdepartmental commission directed by the chairman of the Institute of Nutrition at the Academy of Medical Science, USSR. The tables have appeared in three volumes. The first volume contains nutrient data for 1,592 products as specified in table 2 (SU, column I). The second volume contains approximately 100 nutrients for more than 700 of the food items presented in the first volume. These nutrients include: (1) 18 amino acids; (2) 22 lipids (total, triglycerides, phospholipids, sitosterols) and fatty acids (saturated, monounsaturated, and polyunsaturated); (3) 14 carbohydrates; (4) 17 minerals (8 macroelements and 19 microelements); (5) 13 vitamins.

The third volume contains information on the composition and energy values of the 500 most commonly consumed, ready-to-eat meals and food products. Included are soups, potato and vegetable dishes, cereals, pasta, egg and milk dishes, and fish, meat and poultry dishes, as well as dough, sweet dishes, and beverages.

Eighty percent of the data contained in the first and second volumes of these tables were obtained experimentally by the research institution of the Institute of Nutrition. The remaining data were gathered from the technical literature and verified. All data in the third volume were obtained solely by laboratory analyses. Products are numbered in a decimal classification system (group, subgroup, products). Volumes 1–3 contain an index of products but do not have systematic (Latin) names.

Yugoslavia (YU)

Since 1959, the Institute of Public Health of Croatia has published several editions of food composition tables. The current third edition contains data for almost 500 products, calculated for 100 g e.p. of food derived mainly from compilations. The edition currently in preparation is based on the Institute's own analysis as well as values from the foreign literature. All products will be numbered by a 4-digit code system (10 main groups, subgroup, ordinal number in food subgroups). The food groups of these new tables will be as follows: (1) milk and milk products; (2) meat, game, poultry, fish, and eggs; (3) fat, oils, and their products; (4) legumes, grains, and their products; (5) sugar, honey, and sweets; (6) vegetables and vegetable products; (7) fruit and fruit products; (8) beverages; (9) miscellaneous, and (10) food for special use.

Food Composition Analysis

In addition to the various methods of growing, harvesting, or storing foods, nutrient analysis can lead to discrepancies between values in food tables. Information on analytical methods used for determination of nutrients is a very important factor for interpretation of data and should be referenced in food composition tables [24]. Not all EEFC tables include the analytical methods used to obtain values for nutrients presented. Outlined in table 3 and discussed below are the nutrients included in the tables and methods used to obtain these values.

Energy

Energy values have been calculated based on the amounts of protein, fat, and carbohydrate. Not all tables take into consideration energy from

Table 3. Methods of nutrient analysis used for EEFC tables

Nutrient	Country (reference No.)		
	CS [6]	CS [8]	GDR [9]
Water			
Proteins	crude protein = N × 6.25 for nuts, cocoa, vegetable oils 5.30 soys 5.71 marrow, melon 5.40 rye, oat flour 5.83 wheat flour 5.70 rice 5.95 gelatine 5.55	crude protein	
Amino acids			
Lipids		all substances extractable with suitable solvent	
Fatty acids			
Cholesterol			
Carbohydrates		saccharides, polysaccharides including crude fiber and secondary substances, e.g., organic acids, heteroglycosides, natural coloring matters, tannins	
Sugars and starch			
Fiber	crude fiber	crude fiber	crude fiber
Ash		incineration and annealing	
Minerals			

H [11]	PL [12]	PL [13]	SU [16]
	drying in an oven or under reduced pressure		
crude protein = N × 6.25 for milk and milk products 6.37	crude protein = N × adequate coefficient of [29]		crude protein = N × adequate coefficient from 5.7 to 6.4 [29]
chromatography		ion-exchange chromatography, colorimetric	ion-exchange chromatography
Soxhlet's method	Soxhlet's or Weibull-Stoldt, Schmidt-Badzynski, Grossfeld's methods		mixed solvent extraction
		chromatography	
		colorimetric, chromatography	
from difference	from difference		
		chromatography, enzymatic, colorimetric	unstandardized methods
crude fiber	Hellendoorn's or Southgate's methods		crude fiber
incineration at 500 °C	incineration at 550 °C		incineration at 450–550 °C
macro: flame photometry micro: flame spectro-photometry	flame photometry or atomic absorption spectrophotometry		atomic absorption spectrophotometry; P, Mo, Co, Cr, F – colorimetric; Cl, S, Ba – chemical; Se – fluorometric; B – spectrographic; I – titration; Al, Pb – different methods

alcohol or organic acids. Carbohydrate is calculated by difference. For the most part, energy values are expressed both in kcal and kJ, but different energy factors have been used in their calculation.

Protein and Amino Acids

The crude protein content of a food is obtained by multiplying the total Kjeldahl nitrogen by a designated conversion factor. Different conversion factors have been used to calculate crude protein content in food items that may result in disparate values when comparing food tables. Amino acid composition is usually determined by chromatographic methods (H, PL, SU). In Polish tables, the values for tryptophan have been derived from separate calorimetric analysis.

Fat and Fatty Acids

The values for total lipids were obtained by Soxhlet's method (H, PL) or its modification (PL) and by using methods involving mixed solvent extraction (CS, SU). As with other nutrients, the type of analysis used can affect the value for fat in the tables. For example, the EUROFOODS interlaboratory trial [25] has demonstrated that acid hydrolyses or solvent extraction methodologies are partially responsible for the variation in total fat content. The methods of analysis for fatty acid composition (by chromatography) and cholesterol content (by calorimetric method and chromatography) are given in the Polish food composition tables only.

Carbohydrate

The method for determining carbohydrate content is given in the Hungarian and Polish food composition tables. As a rule, total carbohydrate is calculated by difference, which defines carbohydrate as follows:

Total carbohydrate =
100 – (Protein content + Fat content + Water content + Ash content)

Such values, including unavailable carbohydrates, also are used to calculate energy. This also may contribute to differences between values in food tables.

Sugar

Sugars, including monosaccharides, disaccharides, and starch, are expressed as monosaccharides in Polish tables. The expression of carbohydrates as grams of polymeric starch or grams of equivalent of monosaccharides is a reason for differences between tables. In the Polish tables, values were obtained by chromatographic, calorimetric, and enzymatic methods. The Russian tables use unstandardized methods.

Fiber

Most of the tables give information on crude fiber content (CS, GDR, H, SU). Polish food composition tables have values for dietary fiber obtained by Hellendoorn's or Southgate's methods. For fiber, there are differences in principles and in resulting analytical values among the several methods used. Especially for dietary fiber and available carbohydrates, a standardization of analytical methods is required.

Ash

Ash corresponds to the minerals obtained after incineration of the organic matter at a temperature in a range of 450–550 °C.

Minerals and Trace Elements

Hungarian food composition tables give values for minerals obtained by flame photometry (macro) and by flame spectrophotometry (micro). Flame photometry or atomic absorption spectrophotometry is used to determine the majority of minerals presented in Polish tables. The SU uses different methods for specific minerals but most often uses atomic absorption spectrophotometry.

Vitamins

In Hungarian tables, values for vitamin A content are obtained spectrophotometrically; for the vitamin B group, microbiologically, and for vitamin C, chromatographically. Polish and Russian food composition tables give information on the content of vitamins obtained by different methods: biologically, microbiologically, calorimetrically, or fluorometrically.

Computer-Related Problems

The computer can be used to create a food data bank in two ways: (1) the existing (printed) food tables can be stored and updated on the computer, and (2) new data can be added and stored in the computer, and from these data new computerized tables can be generated.

Use of Computers with Food Composition Tables

Food composition tables can be an effective part of a computerized system only after the following criteria are fulfilled: (1) unification of food coding system; (2) unification of methods of nutrient analysis; (3) unification of computer systems/hardware and software – operation systems, programming languages, transmission – data equipment, etc.

The East European publications on food composition tables differ dramatically. Neither the Council for Mutual Economic Assistance (COMECON) nor the research contact between the national food and nutrition institutes unified the food composition tables. The low level of computerization in these countries contributed to this problem. The tables published in the years 1965–1980 cannot be adapted easily to computers. Only some of those tables use systematic (Latin) names for vegetables, fruits, and fish (BG, PL), and tables from GDR and SU are codified decimally, which helps in locating individual products later.

The Polish tables are coded in such a way that it is difficult to add new products without changing the entire numbering system. From the Hungarian publication [26] it can be expected that the national tables will be computerized in a short time. The new YU tables expected in 1991 are to use the 4-digit code for all food items, which will be very useful for creating a food data bank.

In PL, about 15 years ago, computers were used for designing balanced meals and menus in mass scale feeding systems. The Institute of Human Nutrition of Warsaw Agricultural University established 'Dieta' in 1983. Based on food composition tables (raw food) and the recommended intake of nutrients, the computer system was designed to analyze and compare the nutritive value of consumed foods according to recommendations for a specific population group [3]. Some other research institutes in Poland, for example, the Institute of Food and Nutrition, the Institute of Mother and Child, and the National Institute of Cardiology, adapted their own nutrient data bank to nutritional surveys in research and education. Recently, IBM PCs and compatibles in East European countries have been in common use. Both the use of these computers and the participation in the works of EUROFOODS, INFOODS, and CODATA will become the positive factors for integration according to the headline of EUROFOODS: 'towards compatibility of nutrient data banks in Europe'.

Food Data Bank as a Source of Food Tables

The Czechoslovakian system illustrates the application of the data bank as a source for food tables. During 1976 through 1980, Dr. Strmiska and his group from the Food Research Institute, Bratislava, established and developed a food data base called Food Section in the Central Data Bank. In 1981, they presented detailed food tables in the form of outputs developed from a comparatively large number of stored data. These food tables, printed on approximately 15,000 pages, provided information on the composition and properties of about 700 Czech and 500 foreign food products [27].

The second phase of development of the Food Section started in 1981. Input data serve as the basis of development of Food Section's automated structure and functional capabilities. Basic input data are consistently evaluated and classified according to these principles:

(1) Origin of information; divided and recorded as *home and foreign.*

(2) Object of information, in the given case *food,* that is marked by a 20-positional code, characterizing its relevance to the basic group and relevant subgroup, expressing its biological origin (family, species), stage of ripeness or maturity, relevant anatomic or technologic part, and method of technological processing. The 20-position food code has a logical structure, including a set of basic information on the given object. The 20-position food code is then coupled with the code of the universal system of economic classification officially used in the CSSR and sets up the computer-based system with the direct possibility for controlling economic aspects of the information deposited in Food Section.

(3) Parameter of information, in the given case *parameter of food,* is marked by a 5-positional code and includes the basic information on the given food. These parameters are arranged into seven logical segments: (a) definition of food that also includes the relevant technical-economic norm of quality: (b) data on production of the given food; (c) data on consumption of the given food; (d) data on its physical properties; (e) data on its chemical composition; (f) data on its nutritional value, and (g) relative econometric parameters, especially expenses and costs.

Numerical data on separate parameters are given by statistical information and are supplemented by further data, especially by references (year, author), relevant variety, origin, method of measurement, and supplementary information.

After transferring the input data on magnetic tape, they are processed automatically for the individual nutritional commodities. The number of input data varies, and according to experience it fluctuates between 1,000 and 50,000. In the first stage of the computerized processing, a first-level data base is created, consisting of built-in inputs. The homogeneous first-level data are statistically processed – their mean, maximum, minimum, standard deviation, and coefficient of variation are calculated. In this way, a mixed, second-level data base is created. After these processes, the data of the first-level and second-level data base are algorithmically processed into new and derived data, expressing theoretical relations between the parameters of the given food item. Then a mixed, third-level data base is created.

The different outputs of the Food Section are the following: (1) standardized outputs allow the developed data base to be checked for accuracy and whether the special needs of the working team are served; (2) standardized outputs of statistically processed input data, listing the whole mixed data base (BD23) food tables in three versions: practical, special, and scientific; (3) ad hoc selection reports (listings) according to the needs of the end-user.

Using a relatively small (about 5%) part of the stored data in the Food Section of the Central Data Bank, the first part of the food tables (raw food materials) was published [8]. The second part of the tables containing food products is ready for publication.

Conclusions

Most of the East European countries have their own food composition tables. They are written in national languages only and may be difficult to use in other countries. The number of food items and mode of their classification in food groups are different among tables as are the lists of nutrients included and their arrangement in the tables.

Not all food composition tables give information on the calculation or analytical methods used to obtain values. Using different factors for calculation (energy values, crude protein) or the differences between analytical methodologies (for example, fiber) are some of the reasons for discrepancies.

Unification and an increased compatibility of food composition tables from different countries is necessary, and more work is needed in this field. A useful tool might be the publication, 'Guidelines for the Preparation of Tables of Food Composition' [28], as well as further information exchanges between different countries.

Comparatively, the low level of computerization in East European countries limits the possibility of using these techniques in the preparation of food composition tables. Recently, however, a dramatic increase in the use of computers in these countries has been observed. This, along with the readiness to implement EEFC tables (CS, H, PL) into the regional food data banks, will be the impetus for building a unified international food data bank.

References

1 Rand WM: Infoods: Progress so far and plans for the future. Ann Nutr Metab 1985; 29(suppl 1):46–48.

2 EUROFOODS: Review of food composition tables and nutrient data banks in Europe. Ann Nutr Metab 1985;29(suppl 1):11–45.

3 West CE (ed): Eurofoods: Towards compatibility of nutrient data banks in Europe. Ann Nutr Metab 1985;29(suppl 1):1–72.

4 Arab L, Wittler M, Schettler G: European food composition tables in translation. Berlin, Springer, 1987.

5 Tashev T, Shishkov G: Tables of the composition of Bulgarian food products, ed 2. Sofia, Medicina i fizkultura Publ, 1975.

6 Krondlova-Skopkova M, Smrha O: Tables of nutritive value of foods, ed 3. Praha, Statni Zdravotnicke Nakladatel'stvi, 1965.

7 Kajaba J, Smrha O: Tables of the composition of food products, ed 2. Bratislava, Slovenske Pedagogicke Nakladatel'stvo, 1985.

8 Strmiska F (ed): Food tables. I. Primary foods. Bratislava, Vyskummy Ustav Potravinarsky, 1988.

9 Haenel H (ed): Energy and nutrient content of foods. Food composition tables. Berlin, VEB Verlag Volk und Gesundheit, 1979.

10 Mohr M, Pankrahts F: Short food composition table. Berlin, VEB Verlag Volk und Gesundheit, 1985.

11 Biró G, Lidner K: Food composition tables, ed 11. Budapest, Medicina Könyvkiado, 1988.

12 Piekarska J, Los-Kuczera M: Composition and nutritive value of food products. Part I, ed 4. Warszawa, Panstwowy Zaklad Wydawnictw Lekarskich, 1983.

13 Los-Kuczera M, Piekarska J: Composition and nutritive value of food products. Parts II–VII. Warszawa, Panstwowy Zaklad Wydawnictw Lekarskich, 1988.

14 Piekarska J, Szczygiel A, Los-Kuczera M: Popular tables of nutritive values of food, ed 2. Warszawa, Panstwowy Zaklad Wydawnictw Lekarskich, 1983.

15 Segal R, Segal B, Gheorghe V, et al: The nutritive value of agricultural food products. Bucuresti, Ed. Ceres, 1983.

16 Skurihin JM, Volgarev MN (eds): Chemical composition of food products. Book 1: Composition tables of basic nutrients and energy value of food products, ed 2. Moskva, VO 'Agropromizdat', 1987.

17 Skurihin JM, Volgarev MN (eds): Book 2: Composition tables of amino acids, vitamins, macro- and microelements, organic acids and carbohydrates, ed 2. Moskva, VO 'Agropromizdat', 1987.

18 Skurihin JM, Sodernikov VA: Book 3: Composition tables of basic nutrients and energy value of ready-to-eat meals and culinary products. Moskva, Legkaja i piscevaja promyslennost, 1984.

19 Brodarec A: Tables of composition and nutritive values of food and beverages, ed 4. Zagreb, Zavbod za zastitu zdravlja SR Hrvatske, 1976.

20 Kolodziejska Z, Szczygiel A: Tablice witaminòw. Zdrowie Publiczne 1938;11/12: 1–67.

21 Boas-Fixen MA, Roscoe MH: Tables of the vitamins content of human and animal foods. Nutr Abstr Rev 1938;7:823.

22 Peterson E, Munsell HE: Vitamin content of foods. US Department of Agriculture Publ No 275, 1937.

23 Rudowska-Koprowska J: Tablice wartosci odzywczych produktòw spozywczych. Warszawa, PZWL, 1954.

24 Kunachowicz H, Los-Kuczera M: The development of the work on East European food composition tables. Zyw Czlow Metab 1988;15:135–144.

25 Hollman PCH, Katan MB: Report of the Eurofoods interlaboratory trial 1985 on laboratory procedures as a source of discrepancies between food tables. Wageningen Report 1985;85:67.

26 Varsanyi I: Food composition tables and data bank in Hungary; in Proceedings of the First Conference: Food Data Bank. Modra (Bratislava), May 17–18, 1989.

27 Strmiska F: Food section of the Central Data Bank; in Proceedings of the First Conference: Food Data Bank. Modra (Bratislava), May 17–18, 1989.

28 Southgate DAT: Guidelines for the preparation of tables of food composition. Basel, Karger, 1974.

29 Merrill AL, Watt BK: Energy value of foods – basis and derivation. US Department of Agriculture Handbook No 74, 1955.

30 Los-Kuczera M (ed): Food products. Composition and nutritive value. Warszawa, Institute of Food and Nutrition Publ No 54, 1990.

Jerzy H. Dobrzycki, PhD, Laboratory of Sensory Analysis of Food Quality, Polish Academy of Sciences, Powsinska 61/63, PL-02-903 Warsaw (Poland)

Subject Index